ACADEMIC
SUCCESS
PRESS, INC.

An Award-Winning Math Book

"For Excellence"
– Florida Publishers Group

"Book of the Year"
– The National Association
of Independent Publishers

WINNING AT MATH

*Your Guide to
Learning Mathematics
Through Successful Study Skills*

Paul D. Nolting, Ph.D.
Learning Specialist

ACADEMIC SUCCESS PRESS, INC.

Library of Congress Cataloging-in-Publication Data

Nolting, Paul D., 1951–
    Winning at math : your guide to learning mathematics through
successful study skills / Paul D. Nolting.
      p.   cm.
    Includes index.
    ISBN 0-940287-26-9 (alk. paper)
    1. Mathematics--Study and teaching. I. Title.
QA11.N583   1997
510'.71'-dc21   96-37939
                                     CIP

Winning at Math:
Your Guide to Learning Mathematics
Through Successful Study Skills
by Paul D. Nolting, Ph.D.

New, Expanded and Revised Third Edition, 1997

Copyright © 1997 by Paul D. Nolting, Ph.D.

Published by Academic Success Press, Inc.
Produced by Rainbow Books, Inc.
Cover and Interior Design by Betsy A. Lampé
Illustrations by James B. King

Academic Success Press
P. O. Box 25002 – Box 132
Bradenton, FL 34206

Printed in the United States of America.

Dedication

This book is dedicated to my baby boy, Eric, and to my wife, Vicki.
It is also dedicated to the thousands of students
who are having difficulty learning mathematics
and to the instructors who are teaching them.

Notice to Students

The third edition of *Winning At Math* has been revised to make it easier to read, and it contains more proven math study skills techniques. The chapters have been rearranged, and a new chapter was developed for students with disabilities. This *Winning at Math* edition also focuses on helping students prepare for the math reform's new teaching techniques. Some of the reading improvements are:

- enlarged type,
- more space between lines,
- shorter paragraphs, and
- an easier-to-understand vocabulary.

New topics include:

- the mathematics reform movement,
- the importance of your math attitude,
- understanding your learning style,
- developing your mathematics learning profile,
- finding your best math instructor,
- improving your self-esteem,
- balancing your working and studying time,
- using a calculator while taking notes,
- developing a problem log,
- the revised math note-taking system,
- developing a math glossary,
- using a calculator for homework and tests,
- how to effectively use a math lab or LRC,
- how to become a good collaborative learner,
- applying memory techniques to learning styles,

– developing practice tests,
– using number sense,
– positive self-talk during tests, and
– studying for the final exam.

The new chapter (11) is "Considerations for Students with Disabilities," and its focus is on students with learning disabilities, head injuries and attention deficit disorder. These students often have difficulties with math which are associated with their specific injuries/disorders. Some of the topics included in this new chapter:
– defining LD, TBI and ADD,
– reasons for math learning problems,
– specific learning strategies,
– appropriate accommodations, and
– developing an Individual College Learning/Testing Plan.

The new/revised references include:
– Student Learning-Style Information, and
– Suggestions to Improve Math Study Skills.

The new figures include:
– Figure 1 (Comparing a Traditional Math Classroom to an AMATYC-Standards Classroom),
– Figure 3 (Math-Learning Profile Sheet I),
– Figure 5 (Student Profile Sheet I),
– Figure 11 (Modified 3-Column Note-Taking Sample),
– Figure 14 (Math Lab/LRC Check Sheet),
– Figure 15 (The Stages of Memory),
– Figure 17 (The 12 Myths About Test Anxiety),
– Figure 18 (Note Card Check System), and
– Figure 19 (Individual College Learning/Testing Plan).

Contents

Acknowledgments

Several mathematics instructors gave their time and suggestions to help revise and improve this text. Each person listed is a national expert in a certain area of mathematics. Several of these national experts are part of the Houghton Mifflin Faculty Development team. These individuals either reviewed the old version of *Winning at Math* and made constructive suggestion and/or helped develop new sections.

A special thanks is given to Mr. Daryl Peterson, Director of Houghton Mifflin Faculty Development. Mr. Peterson had the insight to develop national mathematics workshops, which focus on improving mathematics instruction and student's math study skills in a nonselling environment. The Mathematics Faculty Development team, on which I am honored to be included as a member, has three other members: Dr. Michael Hamm, Mr. William Thomas and Mrs. Ginger Asadoorian. Mrs. Asadoorian, of Quinsigamond Community College, made suggestions for this text on number sense. The other team members made additional suggestions. To obtain information on mathematics workshops, call Faculty Development Programs at (800) 856-5727.

Michael Hamm, Ph.D., who is also a member of the Houghton Mifflin Mathematics Faculty Development team, has been teaching developmental mathematics and college-level mathematics for over 15 years at Brookhaven College. Brookhaven College is part of the Dallas County Community College District. Dr. Hamm's expertise is in collaborative learning and graphing calculators. He taught collaborative learning in the Interdisciplinary Studies Program at Brookhaven College in several disciplines. He is a consultant for Texas Instruments. He conducts regional and national workshops on collaborative learning, math study skills, real-world applications and the graphing calculator. His contributions to this revised text are the 10 steps for using the graphing calculator and information on collaborative learning.

William Thomas, Jr., who is part of the Houghton Mifflin Faculty Development team, has a masters degree in mathematics and in education from the University of Toledo. He currently teaches at the University of Toledo's Community and Technical College, for which he has served as Director of Developmental Education. He currently is the Developmental Mathematics Specialist. Mr. Thomas' expertise is in mathematics reform and multicultural learning. Mr. Thomas has given numerous presentations at local, state and national conferences. He has chaired the National Association for Developmental Education's Math Special Professional Interest Network. He is the Vice-Chair of the Developmental Mathematics Committee of American Mathematical Association of Two-Year Colleges. He was on the writing team that formulated and wrote *Crossroads in Mathematics: Standards for Introductory College Mathematics Before Calculus*. His contribution to this revised text is comparing past mathematics instruction to how mathematics will be taught under mathematics reform.

Mrs. Pamela Watkins is a mathematics instructor at Georgia Southern University. She has a BS in Computer Science and a MST in mathematics. She has taught mathematics, including remedial mathematics, and computer science at the college level for 20 years. She has implemented a study-skills course for students with math deficiencies. She has also led numerous workshops on topics ranging from "Using Writing to Teach Mathematics" to "Graphing Calculators: Applications and Lessons." This new edition includes her specific suggestions on math study skills and how take notes and use a calculator at the same time. This text also includes her suggestions on developing a problem log and notebook organization.

Mr. John Hoover has been an instructor of Developmental Mathematics at Southeastern Louisiana University for the last 10 years. Mr. Hoover, along with his cohorts at SLU, are pioneers in the mathematics reform movement and the use of technology in the classroom. Mr. Hoover has a strong interest in the research and the study of math anxiety. He recently completed a study on math anxiety, implementing stress management and study skills along with math instruction in the classroom. *Winning at Math* was one of the major sources used in the study. The study, which had positive results, is being published. Mr. Hoover's contributions to this revised text include suggestions and ideas relating to math skills, study skills, test taking, math reform and technology.

Ms. Margie Stocking is a mathematics instructor in the nationally acclaimed Special Services Program for students with learning disabilities at West Virginia Wesleyan College. Mrs. Stocking holds a BA in Secondary Education Mathematics and a MA in Education. She has conducted workshops for the National Association of Developmental Education and the West Virginia Association of Developmental Education. She attended the Kellogg Institute at Appalachian State University. Her special project was using mathematics study skills in the mathematics classroom to improve the learning of students with learning disabilities and attention deficit disorder. Her contributions to this revised text were study-skills suggestions for students with attention deficit disorder.

Mr. Bill Reineke is Supervisor of the Mathematics Lab at Manatee Community College, Bradenton, Florida. Mr. Reineke is an expert in utilizing mathematics labs as a positive learning environment. He helped develop both the suggestion for using mathematics labs and the Mathematics Lab/LRC Check Sheet.

A final thanks to Dr. Laura Wiggins who is an expert in study skills and an excellent editor. She assisted in proofreading the entire text and made editing suggestions.

Preface

Math is one of the most difficult subjects in college. Every student must pass math to graduate. In fact, some college and universities require students to take up to four math classes to graduate. Too many students are intimidated by math because they have a poor high school math background or are starting college after many years away from school.

What kind of assistance do most students want? Students want tips and procedures they can easily use to help them improve their math grades. The math study suggestions, however, have to be based on educational research and be statistically proven to increase students' learning and grades. *Winning at Math* is the only math study skills book that has statistical evidence demonstrating an improvement in students' ability to learn math and to make better grades.

The study of math study skills, anxiety reduction and test-taking procedures provided in *Winning at Math* are based on Learning Specialist Dr. Paul D. Nolting's 15 years of research. The research has been with students who had difficulty learning math at college and universities throughout the United States. These techniques are effective with both students who are taking math for the first time or those who have previously failed math. In fact, the evidence also suggests grade improvement in other non-math courses , too. These techniques can definitely work for you!

Introduction

Studying and learning math is different from other courses. For this reason, many students make "A's" and "B's" in all other courses but have difficulty passing math.

By using the suggested study procedures in *Winning at Math*, you will be able to improve your math comprehension and make better grades. If you have previously failed math, *Winning at Math* can greatly increase your chances of passing.

The third edition of *Winning at Math* has a special focus on math reform and future requirements in the math classroom. Mathematicians have developed new instructional processes to improve the nation's math literacy for all students, including students with disabilities and minority students.

This is not a new type of math but a reform in math standards relating to teaching techniques and testing. The American Mathematical Association of Two-Year Colleges (AMATYC) has published a book which is the basis of math reform for colleges and universities.

The math reform proposes a new role for both instructor and

student. The classroom atmosphere will change from instructor-centered to student-centered and from lecture to more discussion of math.

Collaborative learning will be part of the classroom activities with less emphasis on rote memorization and more interest on concepts. You need to know how math reform will affect your classroom, study skills and taking tests.

You can win with *Winning at Math* by studying it on your own, using it as part of a traditional course or as independent study program through a math or Learning Resource Center. Combined with the cassette tape, *How to Reduce Test Anxiety*, *Winning at Math* becomes even more effective. This combination provides for a more complete math-study program.

To discover your math-learning strengths and weaknesses, complete the Math study skills Evaluation, pp. 23-30. The self-scoring section will suggest the exact pages in *Winning at Math* you should first read. For a more complete diagnosis of your math study-skills problems and a personalized, printed prescription for success, use the *Wining at Math: Study Skills Computer Evaluation Software II*.

After reading each chapter, complete the reading assignments and questions located on the last pages of each chapter. By reading *Winning at Math* and completing the assignments, you can dramatically improve your ability to learn math.

Math Study Skills Evaluation

Read each of the items below. Choose the statement in each group that is true of you. Indicate what you *actually* do rather than what you *should* do by circling A, B or C. *If this is a library book, please do not write in it*; instead, make an answer sheet and mark it with your answers. *Be Honest.*

1. I:
    A. seldom study math every school day.
    B. often study math every school day.
    C. almost always study math every school day.

2. When I register for a math course, I:
    A. seldom select the instructor who matches my learning style.
    B. often select the instructor who matches my learning style.
    C. almost always select the instructor who matches my learning style.

3. When I am not successful on math tests, I:
    A. seldom blame the instructor.
    B. often blame the instructor.
    C. almost always blame the instructor.

4. I:
    A. seldom study math at least 8 to 12 hours a week.
    B. often study math at least 8 to 12 hours a week.
    C. almost always study math at least 8 to 12 hours a week.

5. Each week, I:
    A. seldom plan the best time to study math.
    B. often plan the best time to study math.
    C. almost always plan the best time to study math.

6. When I take math notes, I:
    A. seldom copy all the steps to a problem.
    B. often copy all the steps to a problem.
    C. almost always copy all the steps to a problem.

7. I:
    A. seldom use an abbreviation system when taking notes.
    B. often use an abbreviation system when taking notes.
    C. almost always use an abbreviation system when taking notes.

8. When I become confused in math class, I:
    A. seldom stop taking notes.
    B. often stop taking notes.
    C. almost always stop taking notes.

9. I:
    A. seldom ask questions in math class.
    B. often ask questions in math class.
    C. almost always ask questions in math class.

10. I:
    A. seldom stop reading the math textbook when I get stuck.
    B. often stop reading the math textbook when I get stuck.
    C. almost always stop reading the math textbook when I get stuck.

11. After reading the math textbook, I:
    A. seldom mentally review what I have read.
    B. often mentally review what I have read.
    C. almost always mentally review what I have read.

12. I:
    A. seldom review class notes or read the textbook assignment before doing my homework.
    B. often review class notes or read the textbook assignment before doing my homework.
    C. almost always review class notes or read the textbook assignment before doing my homework.

13. I:
    A. seldom fall behind in completing math homework assignments.
    B. often fall behind in completing math homework assignments.
    C. almost always fall behind in completing math homework assignments.

14. I:
    A. always properly use my calculator in class, during homework and during tests.
    B. often properly use my calculator in class, during homework and during tests.
    C. seldom properly use my calculator in class, during homework and during tests.

15. There:
    A. seldom are distractions that bother me when I study.
    B. often are distractions that bother me when I study.
    C. almost always are distractions that bother me when I study.

16. I:
    A. seldom do most of my studying the night before the test.
    B. often do most of my studying the night before the test.
    C. almost always do most of my studying the night before the test.

17. To find my best learning resources in the math lab or learning resource center I:
    A. review all of the learning resources.
    B. review some of the learning resources.
    C. review none of the learning resources.

18. I:
    A. am unaware of the characteristics of a good group member.
    B. know some of the characteristics of a good group member.
    C. know all of the characteristics of a good group member.

19. I:
    A. seldom develop memory techniques to remember math concepts.
    B. often develop memory techniques to remember math concepts.
    C. almost always develop memory techniques to remember math concepts.

20. When I have difficulty understanding the math topic, I:
    A. seldom go to the instructor or tutor.
    B. often go to the instructor or tutor.
    C. almost always go to the instructor or tutor.

21. I:
    A. always use good number sense during a test.
    B. often use good number sense during a test.
    C. seldom use good number sense during a test.

22. I:
    A. seldom become anxious and forget important concepts during a math test.
    B. often become anxious and forget important concepts during a math test.
    C. almost always become anxious and forget important concepts during a math test.

23. When taking a math test, I:
    A. seldom start on the first problem and work the remaining problems in their numbered order.
    B. often start on the first problem and work the remaining problems in their numbered order.
    C. almost always start on the first problem and work the remaining problems in their numbered order.

24. Even when time permits, I:
    A. seldom check over my test answers.
    B. often check over my test answers.
    C. almost always check over my test answers.

25. When my math test is returned, I:
    A. seldom analyze the test errors.
    B. often analyze the test errors.
    C. almost always analyze the test errors.

Math Test Scoring for Study Skills

Put the correct amount of points for each item in Section A and Section B to obtain your score. The order of the items is different for Section A and Section B.

Section A — *Point Value Each Statement*

Items	Answer A (1 point)	Answer B (2 points)	Answer C (4 points)	Pages to Read First
1.	_____	_____	_____	32-36
2.	_____	_____	_____	43
4.	_____	_____	_____	98-104
5.	_____	_____	_____	105-107
6.	_____	_____	_____	116, 126
7.	_____	_____	_____	117-18
9.	_____	_____	_____	126
11.	_____	_____	_____	138
12.	_____	_____	_____	140-41
18.	_____	_____	_____	173-76
19.	_____	_____	_____	188-95
20.	_____	_____	_____	196
24.	_____	_____	_____	225-26
25.	_____	_____	_____	226-33

Total   _____ + _____ + _____ = _____

Section A

Section B — *Point Value Each Statement*

Items	Answer A *(4 points)*	Answer B *(2 points)*	Answer C *(1 point)*	Pages to Read First
3.	_____	_____	_____	84-86
8.	_____	_____	_____	119-22
10.	_____	_____	_____	133-38
13.	_____	_____	_____	143-44
14.	_____	_____	_____	144-47
15.	_____	_____	_____	160-62
16.	_____	_____	_____	164-65
17	_____	_____	_____	166-73
21.	_____	_____	_____	197-99
22.	_____	_____	_____	203-16
23.	_____	_____	_____	222-26

Total _____ + _____ + _____ = _____

Section B

_____ + _____ = _____

Section A   Section B   Grand Total

— A score of 70 or below means you have poor math study skills.
— A score of between 70 and 80 means that you have good study skills, but you can improve.
— A score of 90 means you have excellent math study skills.

Chapter 1

What You Need to Know to Study Math

Mathematics (math) courses are not like other college courses. Because they are different, they require different study procedures. Passing most of your other college courses requires only that you read, understand and recall the subject material. To pass math, however, an extra step is required: You must use the information you have learned to correctly solve math problems.

Learning general study skills can help you pass most of your college courses – except math. Special math study skills are needed to help you learn more and get better grades in math. In this chapter you will find out:

— why learning math is different from learning other subjects,

— what the differences are between high school and college math,

— why your first math test is very important, and

— how new changes in the teaching of math will help you learn.

Why learning math is different from learning other subjects

In a math course, you must be able to do three things:

1. *understand* the material,
2. *process* the material, and
3. *apply* what you have learned to correctly solve a problem.

Of these three tasks, applying what you have learned to correctly solve a problem is the hardest.

Examples: Political science courses require that you learn about politics and public service. But your instructor will not make you run for Governor to pass the course. Psychology courses require you to understand the concepts of different psychology theories. But you will not have to help a patient overcome depression to pass the course. In math, however, you must be able to correctly solve problems to pass the course.

Sequential Learning Pattern

Another reason learning math is different from learning other subjects is it follows a sequential learning pattern. "Sequential learning pattern" simply means that the material learned on one day is used the next day and the next day, and so forth. This building-block approach to learning math is the reason why it is difficult to catch up when you get behind. *All* building blocks must be included to be successful in learning math.

You can compare learning math to building a house. Like a house, which must be built foundation first, walls second and roof last, math must be learned in a specific order. Just as you cannot build a house roof first, you cannot learn complex problems without first learning simple ones.

Unlike other subjects, you cannot forget the material after a math test.

Example: In a history class, if you study for Chapters 1 and 2, and do not understand Chapter 3, and end up studying for and having a test on Chapter 4, you *could* pass. Understanding Chapter 4 in history is not totally based on comprehending Chapters 1, 2 or 3. To succeed in math, however, each previous chapter has to be completely understood before continuing to the next chapter.

Sequential learning affects studying for *tests* in math, as well. If you study Chapter 1 and understand it, study Chapter 2 and understand it, and study Chapter 3 and *do not* understand it, then when you have a test on Chapter 4, you are not going to understand it either, and you will probably not do well on the test.

Remember: To learn the new math material for the test on Chapter 5, you must first go back and learn the material in Chapter 4. This means you will have to go back and learn Chapter 4 while learning Chapter 5. (The best of us can fall behind under these circumstances.) However, if you do not understand the material in Chapter 4, you will not understand the material in Chapter 5 either, and you will fail the test on Chapter 5. This is why the sequential learning of math concepts if so important.

The sequential learning pattern also affects the following:

— your previous math course grade,
— your math placement tests scores, and
— the time elapsed since your last math course.

Sequential learning applies to how much math knowledge you have at the beginning of your course. Students who forgot or never acquired the necessary skills from their previous math course will have difficulty with their current math course. If you do not remember what you learned in your last math course, you will have to relearn the math concepts from the previous course as well as the new material for the current course. In most other courses, such as the humanities, previous course knowledge is not required. However, in math you must remember what the last course taught you so that you are prepared for the current course. Measuring previous course knowledge will be explained in Chapter 2, "How to Discover Your Math-Learning Strengths & Weaknesses."

Sequential learning also affects your math placement test scores. If you barely scored high enough to be placed into a math course, then you will have math-learning gaps. Learning problems will occur when new math material is based on one of your learning gaps. The age of the placement test score also affects sequential learning. Placement test scores are designed to measure your *current* math knowledge and are to be used immediately.

Sequential learning is interrupted if math courses are taken irregularly. They are designed to be taken one after another. By taking math courses each semester, without semester breaks in between courses, you are less likely to forget the concepts required for the next course.

Math as a Foreign Language

Another way to understand studying for math is to consider it a foreign language. Looking at math as a foreign language can improve your study procedures. In the case of a foreign language, if you do not practice the language, what happens? You forget it. If you do not practice math, what happens? You are likely to forget it, too. Students who excel in a foreign language must practice it *at least* every other day. The same study skills apply to math, because it is considered a foreign language.

Like a foreign language, math has unfamiliar vocabulary words or terms which must be put in sentences called *expressions* or *equations*. Understanding and solving a math equation is similar to speaking and understanding a sentence in a foreign language.

Example: Math sentences use symbols (which are actually spoken words) in them, such as

"=" (for which you *say*, "equal"),
"-" (for which you *say*, "less"), and
"a" (for which you *say*, "unknown").

Learning *how* to speak math as a language is the key to math success. Currently, most universities consider computer and statistics (a form of math) courses as foreign languages. Some universities have now gone so far as to actually classify math as foreign language.

Math — The Unpopular Subject

Unfortunately, math is not a popular topic. You do not hear the nightly news anchor on television talking in math formulas. He/she talks about major events to which we can relate politically, geographically and historically. Through television — the greatest of learning tools — we learn English, humanities, speech, social studies, and natural sciences, but we do not learn (or even hear about) math.

YOU DO NOT HEAR THE NIGHTLY NEWS ANCHOR ON TELEVISION TALKING IN MATH FORMULAS.

Math concepts are not constantly reinforced in our everyday lives like English, geography, history or other subject areas. Math has to be learned independently. Since it is not reinforced in our everyday lives, it requires more study time.

Math as a Skill Subject

Math is a *skill subject*, which means you have to practice, actively, the skills involved to master it. Learning math is similar to learning to play a sport, learning to play a musical instrument or learning auto mechanics skills. You can listen and watch your coach or instructor all day, but unless you *practice* those skills yourself, you will not learn.

Examples: In basketball, the way to improve your free throw is to *see and to understand* the correct shooting form and then to practice the shots yourself. Practicing the shots improves your free-throwing percentage. However, if you simply listened to your coach describe the correct form and saw him demonstrate it, but you did not *practice* the correct form yourself, you would not increase your shooting percentage.

Suppose you wanted to play the piano, and you hired the best available piano instructor. You would sit next to your instructor on the piano bench and watch the instructor demonstrate beginning piano-playing techniques. Since your piano instructor is the best available, you will see and understand how to place your hand on the keys and play. But what does it take to learn to play the piano? You have to place your hands on the keys and *practice*.

Math works the same way. You can go to class, listen to your instructor, watch the instructor demonstrate skills, and you can understand everything that is said (and feel that you are quite capable of solving the problems). However, if you leave the class *and do not practice* – by working and successfully solving the problems – you will not learn math.

Many of your other college courses can be learned by methods other than practicing. In social studies, for example, information can be learned by listening to your instructor, taking good notes, and participating in class discussions. Many first-year college students mistakenly believe that math can be learned the same way.

Remember: Math is different. If you want to learn math, *you must practice*. Practice not only means doing your homework but it also means spending the time it takes to understand the reasons for each step in each problem.

A Bad Math "Attitude"

Students' attitudes about learning math are different from their attitudes about learning their other subjects. Many students who have had poor past experiences with math do not like math and have a bad attitude about learning it. In fact, some students actually *hate* math, even though these same students have positive experiences with their other college subjects and look forward to going to class.

Society, as a whole, reinforces students' negative attitudes about math. It has become socially acceptable not to do well in math. This negative attitude has become evident even in popular comic strips, such as *Peanuts*. The underlying message is that math should be feared and hated, and that it is all right not to learn math.

The result of this "popular" attitude toward math could reinforce your belief that it is all right to fail math. Such a belief is constantly being reinforced by others in many students' lives.

Example: Students frequently receive sympathy from others when they fail in math, while the same people will criticize the student for failing other courses. Students hear such statements as, "Don't worry about your grade in math; *everybody* flunks math. But your *English* scores really stink. You'd better get on the stick and work on doing better in English." Thinking that it is all right to fail math can lead to missing class and not completing your homework.

The bad math attitude is not a major problem, however. Many students who hate math pass it anyway, just as many students who hate history still pass it. The major problem concerning the bad math attitude is how you use this attitude. If a bad math attitude leads to poor class attendance, poor concentration and poor study skills, then you have a bad math attitude *problem*. You need to read Chapter 3, "How to Take Control and Learn Math," as soon as possible to understand the reasons for these behaviors.

Remember: Passing math is your goal, regardless of your attitude.

What the differences are between high school and college math

Math, as a college-level course, is almost two to three times as difficult as high school-level math courses. There are many reasons for the increased difficulty: Course class-time allowance, the amount of material covered in a course, the length of a course, and the college grading system.

The first important difference between high school and college math courses is the length of time devoted to instruction each week. College math instruction, for the fall and spring semesters, has been cut to three hours per week; high school math instruction is provided five hours per week. Additionally, college courses cover twice the material in the same time frame as do high school courses. What is learned in one year of high school math is learned in one semester (four months) of college math.

Simply put, in college math courses you are receiving less instructional time per week and covering twice the ground per course as you were in high school math courses. The responsibility for learning in college is the student's. As a result, most of your learning (and *practicing*) will have to occur outside of the college classroom.

College Summer Semester Versus Fall or Spring Semesters and the Difference Between Night and Day

College math courses taught during summer semesters are more difficult than those taught during fall or spring. Further, math taught during night courses are more difficult than math taught during day courses.

Students attending a six-week summer math session must learn the information — and master the skills — two-and-a-half times as fast as students attending regular, full-semester math sessions. Though you receive the same amount of instructional classroom time, there is less time to understand *and practice the skills* between class sessions.

Summer semester classes are usually two hours per day, four days per week (nighttime summer classes are four hours per night, two nights per week).

> **Example:** If you do not understand the lecture on Monday, then you have only Monday night to learn the material before progressing to more difficult material on Tuesday. During a night course, you have to learn and understand the material before the break; after the break, you will move on to the more difficult material — *that night.*

Since math is a sequential learning experience, where every building block must be understood before proceeding to the next block, you can quickly fall behind, and you may never catch up. In fact, some students become lost during the first half of a math lecture and, therefore, never understand the rest of the lecture (this can happen during just one session of night class). This is called "kamikaze" math since most students do not survive summer courses.

If you *must* take a summer math course, take a 10- or 12-week *daytime* session so that you will have more time to process the material between classes.

Course Grading System

The course grading system for math is different in college than in high school.

> **Example:** While in high school, if you make a "D" or borderline "D/F," the teacher more than likely will give you a "D," and you may continue to the next course. However, in some college math courses, a "D" is not considered a passing grade, or, if a "D" is made, the course will not count toward graduation.

College instructors are more likely to give the grade of "N" (no grade), "W" (withdrawal from class), or "F" for barely knowing the material. This is because the instructor knows you will be unable to pass the next course if you barely know the current one.

Most colleges require students to pass two college-level algebra courses to graduate. In most high schools, you may graduate by passing one to three math courses. In some college degree programs, you may even have to take four math courses, and make a "C" in all of them, to graduate.

The grading systems for math courses are very precise compared to the systems in English or humanities courses.

> **Example:** In a math course, if you have a 79 percent average and 80 percent is a "B," you will get a "C" in the course. On the other hand, if you made a 79 percent in English, you may be able to talk to your instructor and do extra credit work to earn a "B."

Since math is an exact science and is not as subjective as English, do not expect to talk your math instructor into extra work to earn a better grade. In college, there usually is not a grade given for "daily work," as is often offered in high school. In fact, *your test scores may be the only grades which will count toward your final grade.* Therefore, you should not assume that you will be able to "make up" for a bad test score.

The Ordering of College Math Courses

College math courses should be taken, *in order*, from the fall semester to the spring semester. If at all possible, avoid taking math courses from the spring to fall semesters. There is less time between the fall and spring semester for you to forget the information. During the summer break, you are more likely to forget important concepts required for the next course and, therefore, experience greater difficulty.

Why your first math test
is very important

Making a high grade on the first major college math test is more important than making a high grade on the first major test in other college subjects. The first major math test is the easiest and, most often, is the one the student is least prepared for.

Beginning college students often feel that the first major math test is mainly a review and that they can make a "B" or "C" without much study. These students are overlooking an excellent opportunity to make an "A" on the easiest major math test of the semester. (Do not forget that this test counts the same as the more difficult remaining math tests.)

At the end of the semester, these students sometimes do not pass the math course or do not make an "A" because of their first major test grade. In other words, the first math test score was not high enough to "pull up" a low test score on one of the remaining major tests.

Studying hard for the first major college math test and obtaining an "A" offers you several advantages:

— A high score on the first test can compensate for a low score on a more difficult fourth or fifth math test. All major tests have equal value in the final grade calculations.

— A high score on the first test can provide assurance that you have learned the basic math skills required to pass the course. This means you will not have to spend time relearning the misunderstood material covered on the first major test while learning new material for the next test.

— A high score on the first test can motivate you to do well. Improved motivation can cause you to increase your math study time, which will allow you to master the material.

— A high score on the first test can improve your confidence for higher test scores. With more confidence, you are more likely to work harder on the difficult math homework assignments, which will increase your chances of doing well in the course.

Selecting a College Math Instructor

College math instructors treat students differently from high school math instructors. College math instructors do much less hand-holding than do high school teachers. High school math teachers will frequently warn you about your grades and offer help or makeup work. College math instructors will expect *you* to keep up with how well or poorly you are doing. You must take responsibility for your own success and make an appointment to seek help from your instructor.

Sometimes, and due to the increase in the number of college math courses offered in the curriculum, there are more part-time math instructors than there are full-time math instructors. This problem can restrict student and instructor interaction.

Full-time instructors have regular office hours and are required to be available to help students a certain number of hours per week, either in their offices or in the math lab. Part-time instructors, on the other hand, are only required to teach their math courses; they do not have to meet students after class, even though some part-time instructors will provide this service.

Since math students usually need more instructor assistance after class than do students in other subjects, having a part-time math instructor could require you to find another source of after-class course help. When choosing your classes, try to select a math instructor who is a full-time instructor. You can ask your counselor to help you determine which instructors are full-time and which are part-time

How new changes
in the teaching of math
will help you learn

During the past few years, a major change in the teaching of math has taken place. Mathematicians have developed new teaching techniques with the goal of improving the nation's math literacy. This is not a new type of math, but it is an improvement in the national standards which tell teachers how math should be taught and how students should be tested.

Changes to Expect

The American Math Association of Two-Year Colleges (AMATYC) has published a book titled, *Crossroads In Math: Standards for Introductory College Math Before Calculus*. This book is the basis for the changes in the way colleges and universities are expected to teach math courses. It proposes a new role for both instructor and student. You need to know how math reform will affect *your* classroom.

The college-level math classroom will change in two major ways: It will change from instructor-centered learning to more student-centered learning, and it will change from mostly lecture to more discussion of math. In the past, math classes usually involved the instructor giving lectures and illustrating skills on the chalkboard. Unless there were questions posed to the instructor, a math student was rarely heard from – the instructor was "center stage."

The new changes will mean that the classroom will be a little more like high school was (in the "give and take" between instructor and students), and it will be more student-friendly. Instead of having to listen to endless, long lectures from the instructor about math, you will be asked to participate in teacher/student discussions about math.

Mathematicians believe that these and other changes in the teaching of math will help you better learn and remember what you are taught.

Along with an improved student role in the classroom, college math classes will place more emphasis on learning math concepts and less on math memorization. Additionally, the graphing calculator will be required in most classrooms, where a greater demand will be placed on how to use math skills rather than on basic skill concepts. Since the graphing calculator is designed to perform most math skills, there will be less instruction on skills to be done from memory.

The main focus of changes in the teaching of math will be to include more "real life" problems – especially word problems. Such problems will teach the student how to use math in everyday life. Changes in testing may consist of group tests, individual projects and individual tests with explanations of answers.

As a result of the changes in the teaching of math, you will need to be ready to learn math in a different way. Review Figure 1 (Comparing a Traditional Math Classroom to an AMATYC-Standards Classroom), on the following two pages, to better understand how changes in the teaching of math will affect you.

Figure 1

Comparing a Traditional Math Classroom to an AMATYC-Standards Classroom

Traditional Math Classroom	*AMATYC-Standards Classroom*
The teacher talks to the students. Students listen.	The teacher talks with students. Students talk with each other. Students talk with the teacher.
The teacher tells the concept and gives examples of applying that concept.	The teacher gives students a problem to solve. In solving that problem, students "discover" the concept. The class then discusses the concept.
The teacher shows concepts using only variables.	The teacher exposes students to concepts through problems using numbers, graphs *and* variables.
The teacher uses only the chalkboard and overhead.	The teacher uses chalkboard, overhead, computer, graphing calculator and technology.
The teacher gives students the way to do a problem.	The teacher gives students a chance to experience and do math.
The teacher never uses concrete models.	The teacher uses manipulative techniques to show a concept and gives students problems involving manipulative techniques.
The teacher gives students a multiple-choice or short-answer test that measures what the student has memorized.	The teacher gives students a test that requires the student to explain answers. The test may be a group test.

continued on the next page

Figure 1, continued

Comparing a Traditional Math Classroom to an AMATYC-Standards Classroom

Traditional Math Classroom	*AMATYC-Standards Classroom*
The teacher assumes students know nothing about the subject.	The teacher assumes students know something about the subject, finds out what they know and tailors the instructions to what students know.
Students have no influence in how and what is taught.	Students help the teacher determine what is covered and how it should be covered.
The teacher does not focus on math study skills.	The teacher helps students learn math study skills.
Students work by themselves.	Students work in groups and by themselves.
Students do not write.	Students write journals, papers, portfolios and/or projects.
Students do most work using pencil and paper.	Students do most work using calculator and/or computer.
Students give an answer.	Students give an answer and explain why they got the answer.
Students study math with no connection to anything else.	Students study math and how it is used in other fields of study.
Students solve "simple" problems, which can be solved in a few minutes, using one or two problem-solving strategies.	Students solve "complex" problems, which can be solved over a longer period of time, requiring several problem-solving strategies.

Summary

- The skills required for learning math differ from the skills required for learning other courses.

- Math requires sequential learning, which means one concept builds on the next concept.

- Missing a few major math concepts could cause you to fail the course.

- Thinking of learning math like learning a foreign language or a musical instrument will help you change your math study skills.

- Keeping a positive attitude about math will help you study more efficiently.

- Passing most courses requires only reading, understanding and recalling the subject material.

- When taking a math test, you not only have to understand and recall the material, you have to prove this to the instructor by correctly working the problems.

- In most other subject tests, you can just guess at the answers; you cannot guess at the answers in math tests, because the answers on a math test must be precise and exact.

- Math courses are even more difficult because a grade of "C" or better is usually required to take the next course or, in some cases, just to pass the current course.

- The grading is exact and, in many cases, you cannot do extra credit work to improve your grade.

- Remember to study hard for your first test because it will be the easiest one and, therefore, the greatest opportunity to get a high grade.

- Changes in the way math is taught are coming soon to your college/university or they have already arrived.

- You need to be prepared to learn math in a different way.

- Make sure you know how to use your calculator.

- Be prepared to work in groups to solve math problems.

    Remember: Learning math takes different skills than learning your other subjects.

Assignment for Chapter 1

1. Why is math considered to have a sequential learning pattern?

2. How is math similar to a foreign language and a skill subject?

3. How do attitudes toward math affect math learning?

4. How are high school and college math courses different?

5. How do summer and fall semester math courses differ?

6. How does your math grading system differ from that of your other courses?

7. How can you prepare for your first math test?

8. What are your math instructor's policies about missing classes and tests?

9. How are changes in the way math is taught affecting your classroom?

Chapter 2

How to Discover Your Math-Learning Strengths & Weaknesses

Math-learning strengths and weaknesses affect students' grades. You need to understand your strengths and weaknesses to improve your math-learning skills.

Just as a mechanic does a diagnostic test on a car before repairing it, you need to do your own testing to learn what you need to improve upon. You do not want the mechanic to charge you for something that does not need repairing nor do you want to work on learning areas that do not need improvement. You want to identify learning areas you need to improve or better understand.

In this chapter you will learn:

— how what you know about math affects your grades,

— how quality of math instruction affects your grades,

— how "affective student characteristics" affect your math grades,

— how to determine your learning style,

— how to develop a "math-learning profile" of your strengths and weaknesses, and

— how to improve your math knowledge.

Areas of math strengths and weaknesses include math knowledge, level of test anxiety, study skills, study attitudes, motivation and test-taking skills. Before we start identifying your math strengths and weaknesses, you need to understand what contributes to math academic success.

Dr. Benjamin Bloom, a famous researcher in the field of educational learning, discovered that your IQ (intelligence) and your "cognitive entry skills" account for 50 percent of your course grade. See Figure 2 (Variables Contributing to Student Academic Achievement) on the next page. Quality of instruction represents 25 percent of your course grade, while "affective student characteristics" reflect the remaining 25 percent of your grade.

— *Intelligence* may be considered, for our purpose, how fast you can learn or relearn math concepts.

— *Cognitive entry skills* refer to how much math you knew before entering your current math course.

— *Quality of instruction* is concerned with the effectiveness of math instructors when presenting material to students in the classroom and math lab. This effectiveness depends on the course textbook, curriculum, teaching style, extra teaching aids (videos, audio cassettes), and other assistance.

— *Affective student characteristics* are characteristics you possess which affect your course grades — excluding how much math you knew before entering your math course. Some of these affective characteristics are anxiety, study skills, study attitudes, self-concept, motivation and test-taking skills.

Figure 2

**Variables Contributing to
Student Academic Achievement**

Cognitive Entry Skills
*(how much math you know before
entering a new math course)*
and
IQ
(how fast you can learn old and new math concepts)

— 50% —

Quality of Instruction
*(Effectiveness of Math
Instructors: course
textbook, teaching style,
extra teaching aids, etc.)*

— 25% —

Affective Characteristics
*(Personality: self-concept,
locus of control,
attitudes, anxiety;
Study Habits)*

— 25% —

*Bloom (1976)

How what you know about math affects your grades

Poor math knowledge can cause low grades. A student placed in a math course that requires a more extensive math background than the student possesses will probably fail that course. Without the correct math background, you will fall behind and never catch up.

Placement Tests and Previous Course Grades

The math you need to know to enroll in a particular math course can be measured by a placement test (ACT, SAT) or is indicated by the grade received in the prerequisite math course. Some students are incorrectly placed in math courses by placement tests. Without the correct math background, you will fall behind and never catch up.

If, by the second class meeting, everything looks like Greek and you do not understand what is being explained, move to a lower-level course. In the lower-level math course you will have a better chance to understand the material and to pass the course. A false evaluation of math ability and knowledge can only lead to frustration, anxiety and failure.

Requests by Students for Higher Placement

Some students try to encourage their instructors to move them to a higher-level math course because they believe they have received an inaccurate placement score. Many students try to avoid noncredit math courses, while other students do not want to repeat a course that they have previously failed. These moves can also lead to failure.

Some older students imagine their math skills are just as good as when they completed their last math course, which was five to 10 years ago. If they have not been practicing their math skills, they are just fooling themselves. Still other students believe they do not need the math skills obtained in a prerequisite math course to pass the next course. This is also incorrect thinking. Research indicates that students who were placed correctly in their prerequisite math course, and who subsequently failed it, will not pass the next math course.

What My Research Shows

I have conducted research on thousands of students who have either convinced their instructors to place them in higher-level math courses or have placed themselves in higher-level math courses. The results? These students failed their math courses many times before realizing they did not possess the prerequisite math knowledge needed to pass the course. Students who, without good reason, talk their instructors into moving them up a course level are setting themselves up to fail.

To be successful in a math course, you must have the appropriate math knowledge. If you think you may have difficulty passing a higher-level math course, you probably do not have an adequate math background. Even if you do pass the math course with a "D" or "C," research indicates you will most likely fail the next higher math course.

It is better to be conservative and pass a lower-level math course with an "A" or "B" instead of making a "C" or "D" in a higher-level math course and failing the next course at a higher level.

This is evident when many students repeat a higher-level math course up to five times before repeating the lower-level math course that was barely passed. After repeating the lower-level math course with an "A" or "B," these students passed their higher-level math course.

How quality of math instruction affects your grades

Quality of instruction accounts for 25 percent of your grade. Quality of instruction includes such things as classroom atmosphere, instructor's teaching style, lab instruction, textbook content and format. All of these "quality" factors can affect your ability to learn in the math classroom.

Interestingly enough, probably the most important "quality" variable is the compatibility of an instructor's teaching style with your learning style. You need to discover your learning style and compare it to the instructional style. Noncompatibility can be best solved by finding an instructor who better matches your learning style. However, if you cannot find an instructor to match your learning style, improving your math study skills and using the math lab/LRC can compensate for most of the mismatch.

Use of the math lab or Learning Resource Center (LRC) can dramatically improve the quality of instruction. With today's new technologies, students are able to select their best learning aids. These learning aids could be video tapes, CD-ROMs, computer programs, math study skills computer programs and math texts.

The quality of tutors is also a major part of the math lab or learning resource center. A low student-to-tutor ratio and trained tutors are essential for good tutorial instruction. Otherwise, the result is just a math study hall with a few helpers.

The math textbook should be up to date with good examples and a solutions manual. This increases the amount of learning from the textbook compared to older, poorly designed texts.

The curriculum design affects the sequence of math courses, which could cause learning problems. Some math courses have gaps between them which cause learning problems for all students.

How "affective student characteristics" affect your math grades

"Affective student characteristics" account for about 25 percent of your grade. These affective characteristics include math study skills, test anxiety, motivation, locus of control, learning style and other variables that affect your personal ability to learn math.

Most students do not have this 25 percent of the grade in their favor. In fact, most students have never been taught *any* study skills, let alone how to study math, specifically. Students also do not know their best learning style, which means they may study ineffectively by using their least effective learning style. However, it is not your fault that you were not taught math study skills or made aware of your learning style.

By working to improve your affective characteristics, you will reap the benefits by better learning math and receiving higher grades. Thousands of students have improved their math affective characteristics and, thereby, their grades by using this text.

How to determine your learning style

Research has shown that matching a student's best learning style with the instructional style improves learning. Research has also shown that students who understand their learning style can improve their learning effectiveness. A learning disadvantage will occur for students who do not know or do not understand their learning style. Students should talk to their instructor or counselor about taking one or more learning style inventories.

Taking Stock of Your Learning Style

There are different types of learning styles assessments. One type of learning style assessment focuses on learning modalities while other assessments focus on cognitive or environmental learning styles. We will focus on learning modalities and cognitive learning styles.

Learning modalities (using your senses)

If it is available to you, take a learning style inventory that measures *learning modalities*. Learning modalities focus on the best way your brain receives information; that is, learning *visually* (seeing), *auditorially* (hearing) or *kinesthetically* (touching).

If possible, take an inventory that offers math-learning modalities such as the *Learning Styles Inventory* (Brown & Cooper, 1978). Other learning-styles inventories can measure learning modalities, but they focus on English or reading-learning modes. Sometimes students' math-learning modes may be different from their language-learning modes.

If you cannot locate a modality-type inventory, then you may want to do a self-assessment on your math-learning modality. Do you learn math best by seeing it, by hearing it or by having hands-on learning experiences?

Hands-on learning means you need to touch or feel things to best learn about them. This is not the most accurate way to determine your learning modality; however, you may use this informal assessment until you have an opportunity to take a more formal modality learning-style inventory.

Cognitive learning styles (processing what you sense)

The second type of learning-styles inventory focuses on *cognitive learning styles*. Cognitive learning styles are based on how you process math information once it is received though a learning modality. In other words, once you have heard, seen or felt the information, how does your brain process it? Based on McCarthy (1981), you can process the information four different ways: Innovative, Analytic, Common sense, or Dynamic. Results of either the *Kolburg Learning Styles Inventory* (1985) or the *McCarthy Inventory* (1987) will determine your cognitive learning style.

If you do not have the opportunity to take a cognitive learning-style inventory, you may want to do a self-assessment of your cognitive learning style. Which of the following best describes you?

Innovative learners solve problems or make decisions by personally relating to them and using their feelings. They learn best by listening and discussing their ideas with others, while searching for the meaning. They are students with the good ideas, and they are innovative. They usually major in social sciences and the psychological helping professions.

Analytic learners look for facts and ask experts for advice when solving problems. They learn by thinking about the concepts and solving problems by using logic. They have more interest in ideas and concepts than in people. They generally major in math and the hard sciences; they enjoy doing research.

Common-sense learners solve problems by knowing how things work. They learn by testing concepts in a practical way and make decisions by applying theories. They want to know practical ways to use theories and concepts. They major in applied sciences, health fields, engineering and technologies.

Dynamic learners solve problems by looking at hidden possibilities and processing information concretely. They learn best by independent self instruction and trial-and-error practice. They are flexible and will take risks. They major in business and sales occupations.

Learning modalities and cognitive learning styles are neither good nor bad. Learning modalities and cognitive styles are concerned with how you best take in information and process it. Different college subject areas are best understood through certain learning modalities and cognitive styles.

Most often, math instructors are visual/abstract learners, and they tend to teach the way they best learn. Visual/abstract learners teach through a visual mode, and they have students process information abstractly. The second most frequent learning modality/style of math instructors is visual/common sense.

> **Remember:** It is important to recognize that students who do not have a math-learning modality/style can still learn math. A mismatch of modality/learning styles between student and instructor can be compensated by good study skills.

How to develop a "math-learning profile" of your strengths and weaknesses

Students have learning strengths and weaknesses that affect math learning and their course grade. They must identify and understand these strengths and weaknesses to know which strengths support them and which weaknesses to improve upon.

> **Examples**: Having good math study skills is a *positive* math-learning characteristic while having high test anxiety is a *negative* math-learning characteristic.

Understanding Math-Learning Profiles

By developing a math-learning profile you will be able to identify your learning strengths and weaknesses. There are two types of math-learning profiles that can be developed.

Figure 3 (Math-Learning Profile Sheet I), pp. 62-63, allows you to rate yourself as having weak, average or good math-learning skills.

Figure 4 (Math-Learning Profile Sheet II), p. 70, looks at specific tests which compare your learning skills to the learning skills of other college students. It uses percentile norms to determine how high or low you score compared to the average college student. The average college student will score at the 50th percentile on these assessments.

By plotting your scores, you will see how you compare to other students. You may be able to use one or both profile systems to measure your learning skills.

Figure 3 (Math-Learning Profile Sheet I), pp. 62-63, measures your previous math study skills, test anxiety, math knowledge, learning styles and math attitude. These measurements are based on both the results of certain surveys and your previous math background. Doing all or some of these personal assessments will provide you with a good learning profile, and they may predict your success in math. Mark your scores on a separate sheet if you borrowed this book from the library.

Evaluate Your Math Study Skills

The first assessment to take is the Math Study Skills Evaluation. The paper-and-pencil version of this survey is in the front of this text, pp. 23-29. This evaluation is self-scoreable and suggests the pages which you should read in *Winning at Math* to immediately improve your math study skills.

An even better way to assess your math study skills is to ask your instructor or counselor about taking the *Winning at Math: Study Skills Computer Evaluation Software* program (Academic Success Press). This computerized math study-skills evaluation will diagnose your specific math-learning problems.

A personalized, printed prescription for success, with brief explanations of your math-learning problems, will be generated from the software program. Read each explanation to understand your learning problems.

To make improvements, follow the suggestions on the printout in the areas in which you receive only one point. These one-point areas indicate your worst math study skills, which should be immediately

Figure 3
Math-Learning Profile Sheet I

Student _____

Math Course _____ Math Instructor _____ Phone _____

1. **Math Study Skills Evaluation (p. 23-29) score _____**

   Excellent _____ Good _____ Poor _____
         (5 points)   (3 points)   (1 point)

2. **Math test anxiety level, based on question 3 of the Math Study Skills Evaluation:**

   Answer "A" _____ Answer "B" _____ Answer "C" _____
         (5 points)   (3 points)   (1 point)

3. **Math knowledge (choose either placement score or last course grade):**

   (Placement score)
   High _____ Middle _____ Low _____
         (5 points)   (3 points)   (1 point)

   (Last course and grade) _____
   "A" _____ "B" _____ "C," "D," "F," "W," "N," "I" _____
    (5 points)   (3 points)   (1 point)

4. **Math attitude:**

   Good _____ Neutral _____ Negative _____
         (5 points)   (3 points)   (1 point)

5. **Years since last math course:**

   1 Year _____ 2-3 Years _____ 4 or more Years _____
         (5 points)   (3 points)   (1 point)

6. **Previous study-skills course or training:**

   Course _____ Study-Skills Training _____ None _____
         (5 points)   (3 points)   (1 point)

Figure 3
Math-Learning Profile Sheet I *(continued)*

7. **Previous/current required reading course:**

   No required college-level reading course _____ (5 points)
   First developmental course _____ (3 points)
   Second developmental course _____ (1 point)

8. **How much control do you have over your life?**

   A lot _____ Some _____ Very little _____
       (5 points)   (3 points)   (1 point)

9. **My learning modality is:**

   Visual _____ Auditory _____ Kinesthetic _____
       (5 points)   (3 points)   (1 point)

10. **My cognitive learning style is:**

    Analytical _____ Common sense _____ Innovative/Dynamic _____
        (5 points)   (3 points)   (1 point)

TOTAL POINTS: _____

SCORING

45+ points — an excellent math-learning profile

40-45 points — a good math-learning profile

40 or below — a poor math-learning profile

improved upon.   Continue on to the other learning suggestions to complete your quick, math study skills training.   For either survey form, evaluation or software, mark your total score on the Math-Learning Profile Sheet I, p. 63.

Measuring Your Math Anxiety Level

The second part of the assessment measures your math anxiety level. Look at question number 3 (p. 24) on the Math Study Skills Evaluation and determine your answer. Your answer should be marked "A," "B" or "C."

Indicate your answer on Section B (p. 29) of the Math Test Scoring for Study Skills. If you took the computer version of the Math Study Skills Evaluation and question number 3 is not included, mark "A" as your answer. Answer "A" indicates low anxiety, answer "B" indicates average anxiety and answer "C" indicates high anxiety. As you know, math anxiety can cause you learning and test-taking problems. Mark your response on question number 2 of the Math Learning-Skills Profile Sheet I, p. 62.

Measuring Previous Math Knowledge

The third part of the assessment measures your previous math knowledge. Your previous math knowledge is an excellent predictor of future math success. In fact, the best predictor of math success is your previous math grade (if taken within one year of your current course).

The measurement of your math knowledge is based on your previous math grade or your math placement test. If your college does not require a placement test, contact the chairperson of the math department to help you with your course placement. Selecting a math course yourself is like playing Russian roulette with five bullets in the chamber. Without help, you will probably commit mathematical suicide.

Visit your counselor or instructor to discuss your math placement test score. Record your placement test score on question 3, p. 62. At the same time, record the placement score required for the next level math course and math level below yours. Ask your counselor or instructor how your placement score relates to the scores for the math course below and above your course.

Find out if your score barely placed you into your course, if your score was in the middle or if your score almost put you into the next course. Mark one of these three outcomes on question 3, p. 62. If your placement score barely put you into your current math course, you will have difficulty learning the math required in that course.

A barely passing placement score means that you are missing the math knowledge you require to be successful, and you have a poor chance for passing the course. A middle placement score indicates that you probably have most of the math knowledge required for the course; however, you still may have difficulty passing the course. A high placement score means you have most of the math knowledge needed to pass and may have a good chance of passing the course. (Still, if you have poor math study skills and high test anxiety you may not pass the course.)

Students who have low placement test scores need to improve their math knowledge as soon as possible. Students with middle placement scores should also consider improving their math knowledge.

Improving your math skills needs to be completed in the first few weeks of the semester or before. If you wait too long, you will become hopelessly lost. Suggestions for improving your math knowledge are included in a later section in this chapter.

If you are taking your second college math course, use your previous math course grade to determine your math knowledge. Your previous course grade is the best predictor of success in your next math course. (However, this is only true if your last math course was taken within the last year.)

Record the name and the grade of your previous math course on question 3, p. 62. Students who made an "A" or "B" in their previous math course have a good chance of passing their next math course. However, students who made a "C" in their previous math course have a poor chance of passing their next math course. Students who made below a "C" or withdrew from the previous math course have virtually no chance of passing their next math course.

Determining Your Math Attitude

Math attitudes can play a major part in being successful in your course. Some students who have poor attitudes toward math may not attend class as often as they should or may procrastinate in doing their homework. Other students avoid math until they *have* to take it. Mark the type of math attitude you have on question 4, p. 62.

The Amount of Time Since Your Last Math Course

Another indicator of math success can be the length of time since your last math course. The longer the time since your last math course, the less likely you will be to pass the current course. The exception to this rule is if you were practicing math while you were not taking math courses. Math is very easily forgotten if not practiced. Indicate on question 5, p. 62, the number of years since your last math course.

Previous Study-Skills Courses or Training

General study-skills training that does not focus on math may help some students in time management, reading techniques, learning styles and overcoming procrastination. If you had a course in study skills, or some study-skills training, mark that area on question 6, p. 62. If you had no study-skills training, indicate that on your profile sheet.

Previous/Current Required Reading Courses

College reading skills are also needed for success in math courses.

This is especially true when you are trying to solve story or word problems.

At community colleges, and some universities, there are three levels of reading courses. Usually, the first two levels are noncredit reading courses, and the third level is for college credit.

If you are enrolled in the lowest reading level, you will have difficulty reading and understanding the math text. Students who are enrolled in the second level may also have difficulty reading the text. Students who are not required to take a reading class or who have college-level reading skills may still have some have difficulty reading the math text.

Remember: Math texts are not written like English or history texts. Even students with college-level reading skills may still experience difficulty understanding their math text.

If you do not know your reading level, ask to take a reading test. If in doubt about your reading level, take a reading class. Indicate the type of reading skills you have on question 7, p. 63.

Your Locus of Control

"Locus of control" is a concept that describes how much control you feel you have over both your life and course grades. Some students feel they have a lot of control over their life and learning, while other students feel they have very little control over their life or grades. Indicate on question 8, p. 63, how much control you feel you have over your math learning and grades.

Your Best Learning Modality

Learning modalities are best described by your preference on receiving lecture information. As discussed, some of these modalities match your math instructor and some do not. Most math instructors teach through a visual mode with the distant second modality being auditory. Mark your best learning modality on question 9, p. 63.

Your Cognitive Learning Style

Cognitive learning styles are best described as how information is processed once it is received by your brain. Most math instructors process information *analytically* and they therefore teach math the same way. Using *common sense* is the second most popular way math instructors cognitively process information.

If you do not process information analytically or through common sense, then you may be at a learning disadvantage. However, improved math study skills can compensate for a mismatch in learning modes/styles. Mark your cognitive learning style on question 10, p. 63.

Evaluating Your Findings

Your profile sheet is an informal way to measure your math strengths, weaknesses and learning modalities/styles. Any area with five points is a strength and an area with one point is a weakness. Review your Math-Learning Profile Sheet I and locate your strengths, weaknesses or mismatches.

Learning modalities/styles are counted the same way but are considered matches and mismatches instead of strengths or weaknesses. Total up your score to determine if you have an overall strong or weak math-learning profile.

Some of the weak areas can be compensated for, while other areas, such as learning styles, can be made more effective. Strategies for improving math knowledge and finding the best instructor will be discussed in the last part of this chapter. Other profile weaknesses will be improved by using the suggestions in this text.

More Specific and Formal Profiling Evaluations

Figure 4 (Math-learning Profile Sheet II), p. 70, is a more formal evaluation of your math-learning skills. Math-Learning Profile Sheet II is optional. To complete it, you will need to take surveys and tests such as the *Math Anxiety Rating Scale* (MARS), *Survey of Study Habits and Attitudes* (SSHA), *Nowicki-Strickland Locus of Control* (NSLC), *Nelson-Denny Reading Test* (NDRT), and the math portion of the *Wide Range Achievement Test* (WRAT III-R).

—   *MARS* is a survey for assessing college math anxiety and will produce a score indicating how much anxiety you have compared to other college students.
—   *SSHA* is a survey measuring your study habits and attitudes compared to other college students who make "A's" in their courses.
—   *NSLC* is an opinion survey that estimates how much control you believe you have over life events. "Internal" students take responsibility for their grades and try to improve their learning skills. "External" students believe they have little control over their grades and blame the school or others for their poor achievement.
—   *NDRT* is a test that measures college vocabulary and reading comprehension, compares your scores to other college students, and gives a percentile rank and grade-level equivalent.
—   *WRAT III-R* (the math portion) measures arithmetic and algebra skills and gives your score as a percentile norm and grade level equivalent.

Consult your instructor, college counselor, learning specialist, Learning Resource Center, reading lab, disabled student services or math lab director for assistance in taking these assessments.

Once you have completed all the surveys and tests, enter the percentile scores on Figure 4 (Math-Learning Profile Sheet II), p. 70.

A completed Math-Learning Profile Sheet II will reveal your learn-

Figure 4
Math-Learning Profile Sheet II

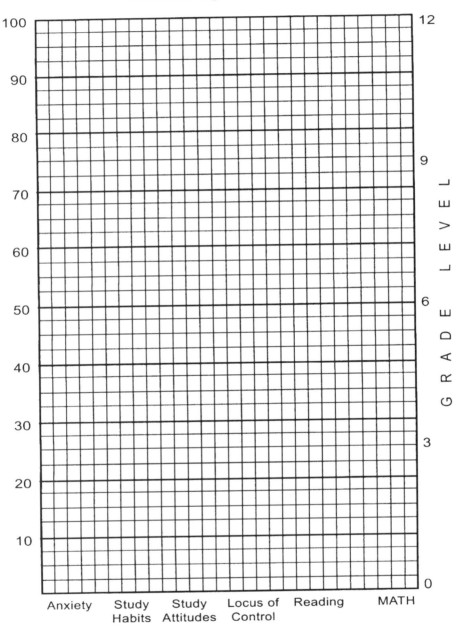

Permission is granted to copy and enlarge Figure 4.

ing strengths and weaknesses as compared to other college students. These assessments give an in-depth picture of your learning strengths and weaknesses. Have your course instructor or counselor explain the meaning of the assessment scores as they relate to improving your math learning and grades.

DISCOVER YOUR STRENGTHS & WEAKNESSES

To better understand Math-Learning Profile Sheet II, look at Figure 5 (Student Profile Sheet I), p. 72, which represents the completed profile of a 30-year-old, married student who works part time and has a family. This student's profile graph contains evaluations of her math anxiety, study habits, study attitudes, locus of control, and reading level scores based on college percentile norms. The math test measured her grade-level equivalent. A high score on the MARS anxiety scale indicates high test anxiety. High scores on the other scales indicate success in those areas.

According to this student's profile, she has extremely high math anxiety, poor study habits, a good study attitude, an internal locus of control, an average reading level, and about a 9th-grade math level. Therefore, she believes that she can make a good grade in math and has a good attitude about school. This will help improve her study skills and decrease her math anxiety. In addition, her reading level and math level are about average and will not present major learning blocks toward improving her math grades.

> **Results:** This student learned how to decrease her math anxiety and improve her study skills while attending my study skills class. She had failed her algebra course, twice, before taking my class. After completing my class, she took the algebra course again and passed with a "B"!

Figure 5
Student Profile Sheet I *(completed)*

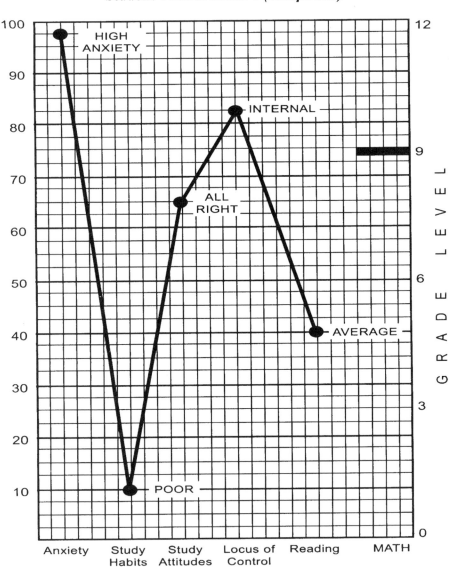

According to this student's profile, she has extremely high math anxiety, poor study habits, a good study attitude, an internal locus of control, an average reading level and about a 9th-grade math level. Therefore, she believes that she can make a good grade in math and has a good attitude about school.

Figure 6
Student Profile Sheet II *(completed)*

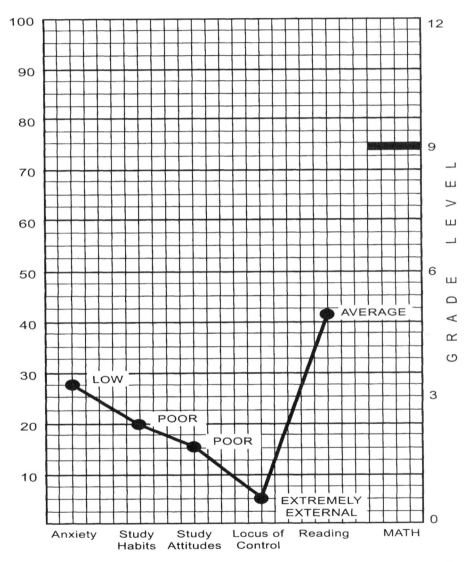

The only two positive characteristics of this student, who has repeatedly failed math, are his low test anxiety and average reading level. He has poor study skills, a poor study attitude and is external in his locus of control. His reading and math skills are average. He must come to believe he can pass math through improving his study habits and attitudes. He needs support from his teacher and counselor to become more "internal" and pass math.

Figure 6 (Student Profile Sheet II), p. 73, represents a student who had a long history of failing math. The only two positive characteristics of this student are his low test anxiety and average reading level. He has poor study skills, a poor study attitude, and is external in his locus of control – all probably due to failing math so many times. His reading and math skills are average. This "external" student must come to believe he can pass math through improving his study habits and attitudes. He also needs support from his teacher and counselor to become more "internal" and pass math.

Results: This student improved his math study skills while attending a math study-skills course; and, by setting up short-term goals, he became more internal. With an increased internal locus of control, he was willing to try some new learning techniques. He passed math that semester.

From these student profiles it is evident that each student has different reasons for being unsuccessful at math. Their problems usually occur in the form of high anxiety, poor study habits, a poor study attitude, or an external locus of control.

How to improve your math knowledge

Instructors always operate on the premise that you finished your previous math course just last week; they do not wait for you to catch up on current material. It does not matter if your previous math course was a month ago or five years ago, instructors expect you to know the previous course material – period.

Review Your Previous Math Course Material and Tests

There are several ways to improve your math knowledge. Review your previous math course material before attending your present math course. Look closely at your final exam to determine your weak areas. Work on your weak areas as soon as possible so they can become building blocks (instead of stumbling blocks) for your current course.

If it has been some time since your last math course, visit the math lab or Learning Resource Center to locate review material. Ask the instructor if there are any computer programs that will assess your math skills to determine your strengths and weaknesses for your course. Review math video tapes on the math course below your level. Also review any computer software designed for the previous math course.

Another way to enhance learning is to review the previous math course text by taking all of the chapter review tests. If you score above 80 percent on one chapter review test, move on to the next chapter. A score below 80 percent means you need to work on that chapter before moving on to the next chapter. Get a tutor to help you with those chapters if you have trouble. Make sure you review all the chapters required in the previous course as soon as possible. If you wait more than two weeks to conclude this exercise, it may be too late to catch up (while learning new material at the same time).

Employ a Tutor

One last way to improve your cognitive entry skills is to employ a private tutor. If you have a history of not doing well in math courses, you may need to start tutorial sessions *the same week class begins*. This will give the tutor a better chance of helping you regain those old math skills.

You still need to work hard to relearn old math skills while continuing to learn the new material. If you wait four to five weeks to employ a tutor, it will probably be too late to catch up and do well or even pass the course.

Remember: Tutorial sessions work best when the sessions begin during the first two weeks of a math course.

Schedule Math Courses "Back to Back"

Another way to maintain appropriate math knowledge is to take each math course "back to back." It is better to take math courses every semester — even if you do not like math — so that you can maintain sequential (linear) learning.

I have known students who have made "B's" or "C's" in a math class, and who then waited six months to a year to take the next math course. Inevitably, many failed. These students did not complete any preparatory math work before the math course and were lost after the second chapter. This is similar to having one semester of Spanish, not speaking it for a year, then visiting Spain and not understanding what is being said.

The only exception to taking math courses "back to back" is taking a six-week "kamikaze" math course (an ultra-condensed version of a regular course), which should be avoided.

If you are one of the unfortunate many who are currently failing a math course, you need to ask yourself, "Am I currently learning any math or just becoming more confused?" If you are learning some math, stay in the course. If you are getting more confused, withdraw from the course. Improve your math knowledge prior to enrolling in a math course during the next semester.

Example: You have withdrawn from a math course after midterm due to low grades. Instead waiting until next semester, attend a math lab or seek a tutor and learn Chapters 1, 2 and 3 *to perfection*. Also use this time to improve your math study skills. You will enter the same math course next semester with excellent math knowledge and study skills. In fact, you can make an "A" on the first test and complete the course with a high grade. Does this sound farfetched? It may, but I know hundreds of students who have used this learning procedure and passed their math course instead of failing it again and again.

Finding Your Best Instructor

Finding an instructor who best matches your learning style can be a difficult task. Your learning style is important; your learning style is how you best acquire information.

> **Example:** Auditory learners do better when hearing the information over and over again instead of carefully reading the information. If an auditory learner is taught by a visual-teaching instructor, who prefers the student read materials on their own and who prefers working problems instead of describing them, the mismatch could cause the student to do worse than if the student were taught by an auditory-teaching instructor.

Most students are placed in their first math course by an academic advisor. Usually academic advisors know who are the most popular and least popular math instructors. However, advisors can be reluctant to discuss teacher popularity. And, unfortunately, students may want the counselor to devise a course schedule based on the student's time limits instead of teacher selection.

To learn who are the best math instructors, ask the academic advisor which math instructor's classes fill up first. This does not place the academic advisor in the position of making a value judgment; neither does it guarantee the best instructor. But it will increase the odds in your favor.

Another manner by which you can acquire a good math instructor is to ask your friends about their current and previous math instructors. However, if a fellow student says an instructor is excellent, make sure your learning style matches your friend's learning style. Ask you friend, "Exactly what makes the instructor so good?" Then compare the answer to how you learn best. If you have a different learning style than your friend, look for another instructor, or ask another friend whose learning style more closely matches your own.

To obtain the most from an instructor, discover your best learning style and match it to the instructor's teaching style. Most learning centers or student personnel offices will have counselors who can measure and explain your learning style. Interview or observe the instructor while the instructor is teaching. This process is time consuming, but it is well worth the effort!

Once you have found your best instructor, do not change. Remain with the same instructor for every math class whenever possible.

Remember: While carefully choosing your best instructor may be a time-consuming inconvenience, it will pay off with higher grades.

Summary

— Math success is primarily based on improving the characteristics that affect your ability to learn math — your *affective* learning characteristics.

— The major affective learning characteristics are study habits, anxiety and control over math.

— There are various ways to improve your math knowledge and math learning.

— Once placed in the appropriate math course, success is based on your ability to learn math.

— Understanding your learning modality and cognitive learning style will improve your learning.

— Learning which affective learning characteristics you need to improve may be accomplished by completing either Figure 3 (Math-Learning Profile Sheet I), p. 62-63, or Figure 4 (Math-Learning Profile Sheet II), p. 70.

— Try to find an instructor who matches your learning style.

— Reading skills are very important in learning math.

— Talk to your instructor or counselor about either of your Math-Learning Profile Sheets.

— Students with high math anxiety need to reach Chapter 9, "How to Reduce Math Test Anxiety," p. 203.

Remember: The first step in becoming a better math student is knowing your learning strengths and weaknesses. Now you can focus on what you need to improve.

Assignment for Chapter 2

1. How does math knowledge affect your math grades ?

2. Using your Math-Learning Profile Sheets, explain your strengths and weaknesses.

3. If possible, take the MARS, SSHA, NDLC, NDRT, WRAT III-R (math only) surveys or tests.

4. If possible, complete your *Math-learning profile Sheet II* and discuss it with your counselor or the course instructor.

5. Explain your cognitive and modality learning styles. How does your learning match up with the typical math instructor's?

6. How can you improve your math knowledge?

7. How can you find the best math instructor for you?

8. What steps are you taking to improve your math-learning weaknesses?

Chapter 3

How to Take Control and Learn Math

You can take control and learn math by understanding your math strengths and weaknesses and by making a commitment to change your learning behaviors.

Students who have problems with math want to improve their learning, but no one has shown them how to change. It is not your fault that you have not been taught how to study math. Even students taking general study-skills courses are often not taught how to study and learn math.

You need to take the responsibility to change your math-learning weaknesses and to use your math-learning strengths. Thousands of students have taken this responsibility and have passed math.

Taking this responsibility requires an internal locus of control. Students with an internal locus of control take the responsibility to change their learning behaviors and do not blame the teacher or the math department for their poor grades. Such students believe that they can change their lives and become a good math students.

Students with an internal locus of control are also aware that, even without knowing it, some part of them may try to sabotage their math learning. This sabotage may be in the form of procrastinating to protect their self-esteem, and procrastination could lead to learned helplessness.

Students with learned helplessness, due to previous math failures, have learned not even to *try* to pass math. By better understanding themselves, such students can guard against the effects of procrastination through the use of successful math study skills. In this chapter you will learn:

— how to develop an internal locus of control,

— how to avoid learned helplessness,

— how to overcome procrastination by defeating the fear of failure, the fear of success and rebellion against authority, and

— how to improve your self-esteem.

How to develop an internal locus of control

The way you can take control over math is by developing an internal locus of control, avoiding learned helplessness and eliminating (or at least decreasing) procrastination.

Defining Locus of Control

Locus of control has to do with the *locus,* or location, in which a student places the control over his/her life; in other words, who or what the student feels controls his/her behavior and grades.

Students who feel that conditions beyond their control prevent them from getting good grades have an *external* locus of control. These students blame instructors, home conditions and money problems for their poor grades and they can do nothing about their problems. In essence, *external* students feel their lives are controlled by *outside forces*, such as fate or the power of other people.

Other students feel they have the power to control their situation, and this power comes from *within*. These *internal* students take responsibility for their success, while external students reject responsibility. Internal students believe that they can overcome most situations, since results depend on their behavior or personal characteristics.

Internal students accept the responsibility for their behavior and realize that studying today will help them pass the math test scheduled for next week. Internal students can delay immediate rewards; they study tonight for a test tomorrow instead of going to a party.

In general, locus of control means that students who are internal will work harder to meet their educational goals than will external students. The internal student can relate today's behavior (e.g., studying, textbook reading) to obtaining a college degree and gainful employment. The external students, on the other hand, cannot connect the behavior of studying today with getting passing grades and future career opportunities. Thus, internals are more oriented toward making high math grades than are externals.

Developing Short- and Long-Term Goals

Externals can change into internals by taking more responsibility for their lives and completing their education. You can take more responsibility by developing and accomplishing short-term goals and long-term goals. Some of your short-term goals can be using the math learning techniques in the next chapters.

Short-term goals are goals developed and accomplished within a day or week.

> **Example:** A short-term goal could be a goal of studying math today between 7:00 p.m. and 9:00 p.m. A long-term goal, for example, could be earning an "A" or "B" in the math course for the semester.

The steps to obtaining your short-term goals or long-term goals must be thought out *and written down*. Keep the written goals in a place where you will see them many times each day. This will help to remind you, frequently, of the goals.

Rewarding yourself after meeting short-term goals increases your internal control by making a strong mental connection between your behavior and the desired reward. Successes in meeting short-term goals lead to greater successes in meeting long-term goals.

How to avoid
learned helplessness

As students become more external (feeling less in control of their lives) they develop *learned helplessness*. Learned helplessness means believing that other people or influences from such things as instructors, poverty or "the system" controls what happens to them. Students who have failed math several times may develop learned helplessness. Some students may ultimately adopt the attitude of "Why try?"

Total lack of motivation to complete math assignments is a good example of learned helplessness. In the past, students may have completed the math assignments but did not get the course grade they had hoped for. This led to the attitude of "Why try?" because they tried several times to be successful in math and still failed the course.

The problem with this thinking is the way these students actually "tried" to pass the math course. Their ineffective learning processes, lack of anxiety reduction and poor test-taking techniques proved to be their math demise. For them, it was like trying to remove a flat tire with a pair of pliers instead of using a tire iron.

This text is your "tire iron." The question is, are you motivated enough to put forth the effort to *use* the tire iron to learn more and make a good grade?

> **Remember:** Students who develop "learned helplessness" *can break this bad habit.* Take responsibility, and you will be on your way to winning at math.

How to overcome procrastination by defeating the fear of failure, the fear of success and rebellion against authority

Procrastination (putting off what needs to be done) is no way to take control over math. Students may procrastinate by not reading the textbook or not doing their homework due to fear of failure, fear of success or rebellion against authority.

Fear of Failure

Some students who fear failure procrastinate to avoid any real assessment of their true ability. By waiting too long to begin work on a paper or studying for a test, real ability is never measured – the rushed paper or unprepared-for test does not reflect what you are *really* capable of accomplishing. Thus, you can never learn the degree of "goodness" or "badness" of your academic ability.

But procrastinators always console themselves after failure, because it does not feel as bad if you only study for a test for two hours and fail it compared to studying for a test for 10 hours and failing.

Another group of students who have fear of failure, and who are thus prone to procrastinate, are the "closet" perfectionists. Perfectionists usually set goals higher than they can realistically reach.

Example: A math student who was failing the course at midterm decided to drop it and to retake it the next semester. She set a goal to make an "A" when the next semester began. After making a "C" on the first major test, she became frustrated at not reaching her goal, and she started procrastinating in her math studies. She fully *expected* to fail the course. This student believed that it was preferable not to try to pass the course if she could not make an "A."

Another example of perfectionism occurs when students attend college after working or being out of school for several years. These students feel they have to make up for lost time by making perfect grades and by graduating faster than younger students. These students look at "B's" as a failure, since their goal is to make 100's on all their tests.

For the older perfectionist, making a low "A" or high "B" on their math test means they are a failure and they want to quit college. Sometimes these goal are so unrealistic — even a genius would fail. Their motto is, "If I can't be perfect, I don't want to try at all."

These students need help to realize that, with their family and/or work responsibilities, making a "B" is all right. They must also learn that finishing a degree later than planned is better than not finishing at all. Being a perfectionist is not related to how high you set the goal; rather, it is *the unrealistic nature* of the goal itself.

Fear of Success

Fear of success means not making an all-out effort toward becoming successful. This is due to the student's fear that "someone" might be hurt or offended by the student's success. Some students believe becoming too successful will lose them friends, lovers or spouses, or that they will feel overwhelmingly guilty for being more successful than their family or close friends.

This "fear of success" can be generalized as "fear of competition" in making good grades. These students do not fear the chance of making low grades when competing. They fear they will not be liked by others if they make high grades.

> **Example:** A math student may fear that by studying too much he will make the highest test grade and set the grading curve. She has more fear that students will not like her due to her high grades than the fear of just making average grades. Such students need to take pride in their learning ability and let the other students take responsibility for their own grades.

Remember: Set the curve — somebody is going to! Grades are *privileged information* between you and the instructor. If a student or family member asks, "How did you do on that test?" Just say, "I passed." Case closed.

Rebellion Against Authority

The third cause of procrastination is the desire to rebel against authority. Some students believe that, by handing in their homework late or by missing the test, they can "get back" at the instructor (whom they may not personally like and whom they may hold responsible for their poor math performance).

These external students usually lack self-esteem and would rather

REBELLING AGAINST AUTHORITY

blame the instructor for their poor grades than take responsibility for completing their homework on time. Rebelling against the instructor gives them a false sense of control over their lives.

However, rebellious students are fulfilling the exact expectations placed on them by their instructors: Becoming academic failures. These students discover, often too late, that they are hurting only themselves.

Procrastination is not a simple issue. Students procrastinate for various reasons. Procrastination, though, is mainly a defense mechanism that protects self-esteem. Most students who procrastinate have poor math grades. By understanding the reasons for procrastination, you can avoid it and become a better math student.

How to improve your self-esteem

Many students experience problems in both their academic and personal lives because they lack self-esteem, that part of our personality which allows us to feel good about ourselves and enjoy our accomplishments. Having self-esteem means that you respect others and have a sense of peace within yourself. Students who have self-esteem also have a "can do" attitude about accomplishing their goals.

Students with poor self-esteem may not put forth as much effort to accomplish their goals. These students can improve their self-esteem by taking responsibility for their feeling, thoughts, abilities and behaviors. Students with poor self-esteem need to change their negative emotional reactions into positive emotional reactions.

Improving self-esteem can be accomplished by changing the negative emotional statements you are telling yourself and by changing the behaviors that can result in poor self-esteem.

Examples: A positive emotional self-statement is, "I accept who I am, and I have the strength to accomplish my goals." An example of changing your behavior is using the suggestions in this text to improve your math study skills.

Remember: Improving your self-esteem will not happen overnight, but you *can* improve it.

10 Ways to Improve Your Self-Esteem

1. When I do well at something, I am going to congratulate myself.

2. I am going to stop procrastinating and blaming others for my problems.

3. When I fail at something, I not going to blame myself but find out how to be more successful the next time.

4. I will not worry about what others think of me.

5. I will do something I like to do, at least one day per week.

6. I will keep a "To Do" list, and I will feel good about myself when marking off each completed item.

7. I will set up short-term and long-term personal/educational goals.

Summary

— Taking control over math means becoming more internal, avoiding learned helplessness and decreasing procrastination.

— You can start internalizing your locus of control by taking the responsibility for following the suggestions in Chapters 4 through 11, while setting and accomplishing realistic short-term academic goals.

— You can reduce your fear of failure by telling yourself that previous math failures were due to poor study skills, not your lack of ability.

— You can avoid learned helplessness by not giving up on making a good grade in math and, if needed, getting help from your instructor and counselor.

— If you have high test anxiety, you can start decreasing procrastination to reduce test anxiety by following the suggestions in Chapter 9, "How to Reduce Math Test Anxiety."

— If you have a math study procrastination problem, you can go directly to Chapter 4, "How to Fit Study Time Into a Busy Schedule," and immediately plan your study time.

— Improve your self-esteem by making positive statements to yourself.

— Do not believe other students when they tell you they "can't do math."

Remember: Students who have followed the suggestions in

this text have significantly improved their math grades as compared to similar students who did not use these suggestions.

Assignment for Chapter 3

1. Meet with your math instructor at least three times during the semester to get feedback on your course progress.

2. Meet with the instructor (or an LRC staff member) for your current math course at least once to discuss what you can do to improve your math grades.

3. On a separate sheet, write an analysis of your reasons for procrastination in math and decide how you are going to overcome them. Keep these posted in an area where you will frequently see them.

4. Meet with your counselor, as needed, to discuss procrastination issues.

5. Write down two short-term goals to improve your internal locus of control.

6. Come up with three appealing rewards for which you would gladly study your math assignments without procrastination.

Chapter 4

How to Fit Study Time Into a Busy Schedule

One of the main problems college students have is managing their time. In high school, the student's time is managed by teachers and parents. In college, students have more activities (work, social, study) but less time to complete these activities (and no teachers or parents handy or willing to schedule their time for them).

When freshman college students are asked to give their number one reason for poor grades, they indicate that they do not have enough time to study. When students are asked how much time they study per week, most do not have any idea.

Students who do not effectively manage their study time may fail math courses. As pointed out in Chapter 1, math requires much *practice* (the same as mastering a sport or musical instrument) for the student to perform well on tests. Therefore, developing a good plan for studying math can be an important key to getting good math grades.

Even mature students entering college have difficulty managing their time. These students have family and work responsibilities which cause time-management problems. These students have a strong desire to study but need help finding the time to study. They, too, must learn to effectively use the little time they have available.

In this chapter you will learn:

— how to develop a study schedule,

— how to prioritize your time,

— how to create a weekly study plan, and

— how to manage working and studying.

How to develop
a study schedule

There are two basic reasons for developing a study schedule: To schedule your study time and to become more efficient at studying.

You need to learn to set aside a certain amount of study time each week. You should emphasize the number of hours, *per week*, you are going to study instead of the number of *daily* study hours.

How many hours do *you* study per week? Ten hours, 15 hours, 20 hours, 30 hours? Without knowing the amount of your study hours, per week, you will not know if you are studying at a productive rate.

> **Example:** If your goal is to make a "B" average, and with studying 15 hours per week you make all "B's" on your tests, then the goal has been met. However, if you study 15 hours per week and make all "D's," then you need to increase your study time.

By monitoring your grades and the number of hours you study per week, you can adjust your study schedule to get the grades you want.

The second reason for developing a study schedule, *efficient* study, means knowing *when* you are supposed to study and when you do *not* have to study. This approach will help keep you from thinking about other things you should be doing when you sit down to study. The reverse is also true. When doing other, more enjoyable things, you will not feel guilty about not studying.

> **Example:** You are at the mall on a Sunday afternoon, shopping for clothes, when you start feeling guilty. You have not started studying for that math test on Monday. If you had created a study schedule, you could have arranged to study for the math test on Saturday and still have been able to enjoy the mall on Sunday.

A study schedule should be set up for two reasons: 1] To determine the amount of study time you need, per week, to get the grades you want, and 2] to set up peak efficient study times.

How to prioritize your time

To develop a study schedule, turn to Figure 7 (Planning Use of Daily Time) on the next page and review it. Use Planning Use of Daily Time as your study schedule (feel free to make enlarged copies of Figure 7).

The best way to begin to develop your study schedule is to fill in all the times you *cannot* study. Do this by following the steps below:

Step 1 – Fill in all your classes by putting code "C" (C=class) in the correct time spaces. For example, if you have an 8:00-9:30 class, draw a line through the center of the 9:00 am box on the study schedule.

Figure 7

Planning Use of Daily Time
(Master Plan)

	Monday	Tuesday	Wednesday	Thursday	Friday	Saturday	Sunday
6:00							
7:00							
8:00							
9:00							
10:00							
11:00							
12:00							
1:00							
2:00							
3:00							
4:00							
5:00							
6:00							
7:00							
8:00							
9:00							
10:00							
11:00							
12:00							

CODES: C=Class, **W**=Work, **E**=Eating, **G**=Grooming, **T**=Tutor, **F**=Family, **CN**=Cleaning, **SL**=Sleep, **SC**=Social Time, **TR**=Travel, **O**=Other, **S**=Study

Grade goal: _____

Permission is granted to copy and enlarge Figure 7.

Step 2 – Fill in the time you work with code "W" (W=work). This may be difficult, since some students' work schedules may change during the week. The best way to predict work time is to base it on the time you worked the previous week unless you are on a rotating shift. Indicate your approximate work times on the study schedule. As your work hours change, revise the study schedule.

> **Remember:** The study schedule is structured around the number of hours a week you plan to study. Realize that while your work times might change every week, your *total* weekly work hours usually remain the same.

Step 3 – Decide the amount of time it takes to eat (E=eat) breakfast, lunch and dinner; this time slot should include both food preparation and clean up. Keep in mind that the amount of time it takes to eat may fluctuate. Include enough time to prepare the meals, eat and clean up afterward. Eating time also includes time spent in the student cafeteria. If you have an 11:00-12:00 or 1:00-2:00 lunch break, you might not eat during the entire time; you could be there both socializing and eating. Still put code "E" in the study schedule, since the main use of your time is for eating.

Step 4 – Include your grooming time (G=grooming). Some grooming activities include taking a bath, washing your hair or other activities that you do to get ready for school, dates or work. Grooming varies from minutes to hours per day for college students. Mark your study schedule with code "G" for the usual amount of time spent on grooming. Remember that more time might be spent on grooming during weekends.

Step 5 – Include your tutor time (T=tutor). This is *not* considered study time. Tutor time is the time during which you ask the tutor questions about the previous homework assignment. If you have a tutor scheduled or meet weekly with your instructor, mark these times with code "T" in the study schedule.

Step 6 – Reserve time for family responsibilities on the study schedule (F=family responsibilities). Some family responsibilities include taking your child on errands, mowing the lawn, grocery shopping and taking out the garbage. Also, if you have arranged to take your children some place every Saturday morning, then put it on the study schedule using code "F."

Step 7 – Figure out how much time is spent on cleaning each week (CN=cleaning). This time can include cleaning your room, house, car, and clothes. Cleaning time usually takes several hours a week. Indicate with code "CN" your cleaning time on the study schedule, and make sure it is adequate for the entire week.

Step 8 – Review your sleep patterns for the week (SL=sleep). Your sleep time will probably be the same from Monday through Friday. On the weekend, you might sleep later during the day and stay up later at night. Be realistic when scheduling your sleep time. If you have been sleeping on Saturday mornings until 10:00 a.m. for the last two or three years, do not plan time at 8:00 a.m. to study.

Step 9 – Figure the amount of weekly social time (SC=social time). Social time includes being with other people, watching TV or going to church. It can be doing nothing at all or going out and having a good time. You need to have some social time during the week or you will burn out, and you will probably drop out of school. You may last only one semester. If you study and work too hard without some relaxation, you will not last the entire school year. Some daily social time is needed, but do not overdo it.

Step 10 – Figure the amount of travel time to and from work (TR= travel time). Travel time could be driving to and from college or riding the subway. If possible, use travel time to listen to audio tapes of your class or to review notes. Travel time may vary during different times of the year.

Step 11 – Recall other time obligations that have not been previously mentioned (O=other). Other time obligations may be aspects of your life which you do not want to share with other people. Review the study schedule for any other time obligations and mark them.

Now, count up all the blank spaces. Each blank space represents one hour.

> **Remember:** You might have several half-blank spaces which each represent one-half hour. Add together the number of blank spaces left and write the total in the oval (lower right-hand corner of the study schedule).

Next, figure how many hours you have to study during the week. The rule of thumb is to study approximately two hours per week for each class hour. If you have 12 real class hours (not counting physical education) per week, you should be studying 20-24 hours per week to make "A's" and "B's." Write the amount of time you *want* to study per week in the square, located in the lower left-hand corner of your study schedule. This is a study "contract" you are making with yourself.

If the number of contracted (square) study hours is less than the number in the oval, then fill in the times you want to study (S=study).

First, fill in the best times to study. If there are any unmarked spaces, use them as backup study time. Now you have a schedule of the best times to study.

On the other hand, if you want to study 15 hours a week and have only 10 hours of space, then you have to make a decision. Go back over your study-schedule codes and locate where you can change some times. If you have a problem locating additional study time, make a priority time list. Take the hours away from the items with the least priority. Complete the study schedule by putting in your best study times.

How to Choose the Grade You Would Like to Make in Math

Determine what grade you want to make in the math course and write the grade on the study schedule. The grade should be an "A," a "B" or a "C." Do not write an "N," "W," "X" or "F" because these grades mean you will not successfully complete the course. Do not write "D" because you may not get credit for the course or be allowed to take the next course. In fact, it is not wise to write "C," either, because most students who make a "C" usually fail the next math course. The grade you selected to make in math is now your goal.

Some developmental math courses are graded on a pass-or-fail basis. Do not take the attitude that all you need to do is pass the course. A passing score for these courses translates into a "C." This means that you have only an average foundation that might not be strong enough to pass the next math course in the sequence. Set your goals to make an average high enough that would be equal to an "A" or a "B."

Currently, you have a study schedule representing the number of hours of study per week. You also have a course grade goal. After you have been in the course for several weeks and get back the results of your first tests, you will know if you are accomplishing your goal. Should you not meet your goal, improve the quality and quantity of your studying or lower your course grade goal.

Choosing a GPA (Grade Point Average) Goal for the Semester

After figuring out what grade you want in the math course, write down on your study schedule the grade point average you want for the semester. Do you want a 4.0 average, (all "A's"), a 3.0 average (all "B's"), a 2.5 average ("B's" and "C's"), or a 2.00 average (all "C's")? Do not choose any average below 2.00; you will not graduate with a lower average. Be realistic when deciding upon an overall grade point average.

How to create
a weekly study plan

By using the information from your completed copy of Figure 7 (Planning Use of Daily Time), you will know which time slots are available for study. You can use this information to both develop an effective study plan for the next week and establish weekly study goals.

Each Sunday, use an enlarged copy of Figure 8 (Weekly Study-Goal Sheet), on the next page, to plan the best use for your study time during the next week. The first priority, when completing Figure 8, is to establish *the best time to study math*. Math should be studied as soon as possible after each class session. Therefore, you should choose for math study the time slots closest to your math-class times.

Be sure to indicate *where* you will study math. Are you going to the math lab to review video tapes or use the software, study in a group at someone's home or work with your tutor in your apartment?

Once you have indicated in the "Weekly Study-Goal Sheet your math study times and locations, you may then fill in your study goals for your other subjects.

This plan becomes especially important and helpful during the weeks of midterm and final exams.

Figure 8 — Weekly Study-Goal Sheet

Subjects	Monday	Tuesday	Wednesday	Thursday	Friday	Saturday	Sunday
MATH							

Example: You have determined, from completing Figure 7 (Planning Use of Daily Time), that you have study time available from 3:00 until 6:00 on Monday, Tuesday, Wednesday and Thursday, and from 1:00 until 6 p.m. on Sunday. You will schedule your math study time before scheduling the study time of other classes.

If your math class is the last class of the day, you should schedule your math study time in the first study-time slot for that day. Therefore, you would choose to begin studying your daily math (what you learned that day) at 3:00 on Monday, Tuesday, Wednesday and Thursday. You will be studying your math at home.

You will mark the box on the math line for Tuesday, Wednesday and Thursday with: 3 p.m., daily work, home.

You have a math midterm exam on the following Monday, so you should schedule your Sunday math study time last. This way, it will be more "fresh" in your mind during the test on Monday. You have chosen to study for this test for two hours in the library with a group of other students.

You will mark the box on the math line for Sunday with: 4 p.m., midterm, library, group.

Now that your math study goals have been set, you may schedule the study time for your other subjects.

How to manage
working and studying

Most college students work and attend college at the same time. Some college students even work full time and try to attend college full time. This can be dangerous because many full-time college students who attempt to work full time become dropouts. Students who work and attend college at the same time must manage their time very closely to balance their work and study schedules. The suggestions on the next page can help you best manage work and study.

Suggestions for Students Who Must Work

• Try to find a job that allows you some opportunity to study.

• Try to arrange to go to work right after class.

• Study as much as you can at school, then go to work. Do not go home first. Take work clothes with you to school, if needed.

• Study during your lunch time or during breaks.

• Try to review your notes during work, if the job allows.

• Record class lectures and play them during your travel to and from work.

• Take only 12 semester hours, which will qualify you as full time for financial aid but will not overburden your time.

• Take one easy course each semester.

• Do not wait until the weekend to do your homework.

• Increase your reading rate and become an efficient studier.

Students can be successful in college while working; however, it takes effective time management along with excellent reading and study techniques. Make sure you are not setting yourself up to be successful at work while failing college. If necessary, try dropping to part time in college and work overtime to obtain enough money to attend college. Also, it is better to "stop out" of college for a semester and work instead of working and attending college if working causes you to have to withdraw or fail your classes.

Summary

— You have now completed a study schedule and a Weekly Study-Goal Sheet indicating both the times to study and the number of study hours per week.

— The number of study hours you contracted with yourself can change based on the grades you want.

— If you do not receive the grade you want in math, or your desired overall average during the semester, then increase the quality and quantity of study time.

— Taking control of your study time can greatly improve your grades.

— If you are working, have a family and are attending college full time, your time management is extremely important.

— You need to develop a plan to study in short spurts of 30 to 45 minutes.

— During that time you can review your notes, do a few problems or even listen to part of an audiotape recording of your class.

— In fact, take the audio cassette tapes and play them in your car while running errands or going to work.

— Full-time parents who are also students need to be creative in developing a workable study time.

Remember: Every little bit of studying helps.

Assignment for Chapter 4

1. Complete Figure 7 (Planning Use of Daily Time).

2. Select an "A," "B," or "C" as a math-course grade goal.

3. Select between a 2.00 and 4.00 grade point average as an overall semester goal.

4. Complete Figure 8 (Weekly Study-Goal Sheet).

5. List five creative ways to study.

Chapter 5

How to Improve Listening and Note-Taking Skills

Listening and note-taking skills in a math class are very important, since most students do not read the math text or have difficulty understanding it. In most of your other classes, if you do not understand the lecture you can read the text and get almost all the information. In the math class, however, the instructor can usually explain the textbook better than the students can read and understand it.

Students who do not have good listening skills or note-taking skills will be at a disadvantage in learning math. Most math understanding takes place in the classroom. Students must learn how to take advantage of learning math in the classroom by becoming effective listeners, calculator users and note-takers.

In this chapter your will learn:

— how to become an effective listener,

— how to become a good note-taker,

— when to take notes,

— the seven steps to math note-taking,

— how to rework your notes, and

— how to correctly use a calculator in class.

How to become
an effective listener

Becoming an effective listener is the foundation for good note-taking. You can become an effective listener using a set of skills, which you can learn and practice. To become an effective listener, you must prepare yourself both physically and mentally.

Sitting in the Golden Triangle

The physical preparation for becoming an effective listener involves *where you sit* in the classroom. Sit in the best area to obtain high grades; that is, in "The Golden Triangle of Success." The Golden Triangle of Success begins with seats in the front row facing the instructor's desk. See Figure 9 (The Golden Triangle of Success) on the next page.

Students seated in this area (especially on the front row) directly face the teacher and are most likely to pay attention to the lecture. This is a great seating location for visual learners. There is also less tendency for them to be distracted by activities outside the classroom or by students making noise within the classroom.

The middle seat in the back row is another point in The Golden Triangle for students to sit, especially those who are auditory (hearing) learners. You can hear the instructor better because the instructor's voice is projected to that point. This means that there is less chance of misunderstanding the instructor, and you can hear well enough to ask appropriate questions.

By sitting in The Golden Triangle of Success, you can force yourself to pay more attention during class and be less distracted by other students. This is very important for math students because math instructors usually go over a point once and continue on to the next

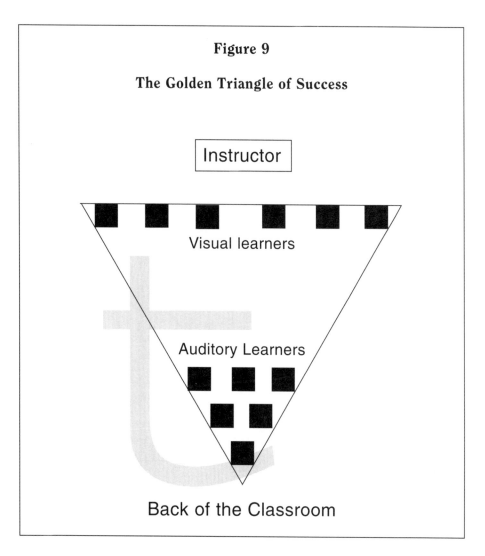

Figure 9

The Golden Triangle of Success

Instructor

Visual learners

Auditory Learners

Back of the Classroom

point. If you miss that point in the lesson, then you could be lost for the remainder of the class.

Warming Up for Math Class

The mental preparation for note-taking involves "warming up" before class begins and becoming an active listener. Just as an athlete must "warm up" before a game begins, you must "warm up" before taking notes. "Warm up" by:

— reviewing the previous day's notes,

— reviewing the reading material,

— reviewing the homework,

— preparing questions, and

— working one or two unassigned homework problems.

This mental "warm up" before the lecture allows you to refresh your memory and prepare pertinent questions, making it easier to learn the new lecture material.

How to Become an Active Listener

Becoming an active listener is the second part of the mental preparation for note-taking.

> **Examples:** Watching the speaker, listening for main ideas and nodding your head or saying to yourself, "I understand," when agreeing with the instructor.

Do not anticipate what the instructor is going to say or immediately judge the instructor's information before the point is made. This will distract you from learning the information.

Expend energy looking for interesting topics in the lecture. When the instructor discusses information that you need to know, immediately repeat it to yourself to begin the learning process.

You can practice this exercise by viewing math video tapes and repeating important information. This is an especially good learning technique for auditory learners.

> **Remember:** Class time is an intense study period that should not be wasted.

Listening and Learning

Some students think listening to the instructor and taking notes is a waste of valuable time. Students too often sit in class and use only a fraction of their learning ability. Class time should be considered a valuable study period where you can listen, take notes and learn at the same time. One way to do this is by memorizing important facts when the instructor is talking about material you already know. Another technique  is to repeat back to yourself the important concepts right after the instructor says them in class. Using class time to learn math is an efficient learning system.

How to become
a good note-taker

Becoming a good note-taker requires two basic strategies. One strategy is to be specific in detail. In other words, *copy* the problems down, step by step. The second strategy is to *understand* the general principles, general concepts and general ideas.

Copying from the Board

While taking math notes, you need to copy each and every step of the problem even though you may already know every step of the problem. While in the classroom, you might understand each step, but a week from then you might not remember how to do the problem unless all the steps were written down. In addition, as you write down each step, you are memorizing it. Make sure to copy every step for each problem written on the board.

There will be times when you will get lost while listening to the lecture. Nevertheless, you should keep taking notes even though you do not understand the problem. This will provide you with a reference point for further study. Put a question mark (?) by those steps which you do not understand; then, after class, review the steps you did not understand with the instructor, your tutor or a fellow student.

Taking Notes

The goal of note-taking is to take the least amount of notes and get the greatest amount of information on your paper. This could be the opposite of what most instructors have told you. Some instructors tell you to take down everything. This is not necessarily a good note-taking system, since it is very difficult to take precise, specific notes while at the same time understanding the instructor.

What you need to develop is a note-taking system in which you write the least amount possible and get the most information down while still understanding what the instructor is saying.

Develop an Abbreviations List

To reduce the amount of written notes, an abbreviation system is needed. An abbreviation system is your own system of reducing long words to shorter versions which you still can understand. By writing less, you can listen more and have a better understanding of the material.

> **Example:** When the instructor starts explaining the commutative property, you need to write it out the first time. After that, use "COM." You should develop abbreviations for all the most commonly used words in math.

Figure 10 (Abbreviations), on the next page, provides a list of abbreviations. Add your own abbreviations to this list. By using abbreviations as much as possible, you can obtain the same meaning from your notes and have more time to listen to the instructor.

When to take notes

To become a better note-taker you must know when to take notes and when not to take notes. The instructor will give cues that indicate what material is important. Some such cues include:

– presenting usual facts or ideas

Figure 10

Abbreviations

E.G. (for example)

CF. (compare, remember in context)

N.B. (note well, this is important)

∴ (therefore)

∵ (because)

⊃ (implies, it follows from this)

> (greater than)

< (less than)

= (equals, is the same)

≠ (does not equal, is not the same)

( ) (parentheses in the margin, around a sentence or group of sentences indicates an important idea)

? (used to indicate you do not understand the material)

0 (a circle around a word may indicate that you are not familiar with it; look it up)

E (marks important materials likely to be used in an exam)

1, 2, 3, 4 (to indicate a series of facts)

D (shows disagreement with statement or passage)

REF (reference)

et al (and others)

bk (book)

p (page)

etc. (and so forth)

V (see)

VS (see above)

SC (namely)

SQ (the following)

Comm. (Commutative)

Dis. (Distributive)

A.P.A. (Associative Property of Addition)

A.I. (Additive Inverse)

I.P.M. (Identity Property of Multiplication)

— writing on the board

— summarizing

— pausing

— repeating statements

— enumerating; such as, "1, 2, 3" or "A, B, C"

— working several examples of the same type of problem on the black-board

— saying, "This is a tricky problem. Most students will miss it." For example, 5/0 is "undefined" instead of "zero."

— saying, "This is the most difficult step in the problem."

— indicating that certain types of problems will be on the test, such as coin- or age-word problems

— explaining bold-print words

You must learn the cues your instructor gives indicating important material. If you are in doubt about the importance of the class material, do not hesitate to ask the instructor about its importance.

While taking notes, you may become confused about math material. At that point, take as many notes as possible, and do not give up on note-taking.

As you take notes on confusing problem steps, skip lines; then go back and fill in information that clarifies your misunderstanding of the steps in question. Ask your tutor or instructor for help with the uncompleted problem steps, and write down the reasons for each step in the space provided.

Another procedure to save time while taking notes is to stop writing complete sentences. Write your main thoughts in phrases. Phrases are easier to jot down and easier to memorize.

The seven steps to math note-taking

The key to effective note-taking is to record the fewest words while retaining the greatest information. As you know, it is very difficult to record notes and, at the same time, fully understand the instructor. The "seven steps to math note-taking" system was developed to decrease the amount of note-taking while at the same time improving math learning.

The seven steps to math note-taking system consists of three major areas. The first area, Steps One through Three, focuses on *recording* your notes. Steps Four through Six focus on *checking* yourself to see how much information is retained. This is done by recalling key words and concepts and putting a check mark by misunderstood information. Recalling information is one of the best learning techniques.

Note-Taking Memory Cues

One of the best math note-taking methods is demonstrated in Figure 11 (Modified Three-Column Note-Taking Sample) on the facing page. To use this effective note-taking system, you need to record a few memory cues as reminders. Label the top space between the notebook ring and the red line, "Key Words." Label the other side of the red line, "Examples." Next, label "Explanations/ Rules" about four inches from the red line. Draw a vertical line between the "Examples" and "Explanations/Rules" sections. Record the same information on

Figure 11
Modified Three-Column Note-Taking Sample

Keywords	Examples	Explanations/Rules
Natural Numbers	1, 2, 3, 4, 5 . . .	You can count them.
Whole Numbers	0, 1, 2, 3, 4, . . .	Natural numbers and zero
Integers	. . . -2, -1, 0, 1, 2 . . .	Negative numbers and whole numbers
-n	-n if n = S then -n = -s opposite of -(-10) = 10 opposite of -(-15)(-15) = -15 opposite of -x = x	Opposite of any number Count the number of signs, even means +, odd means -
Rational	Rational numbers are fractions (¼, ½, ¾)	A/B, B ≠ 0 is rational Division by 0 undefined
Numerator Denominator	numerator/denominator N/D	Numerator on top Denominator on bottom (D = Down)
Terminating Decimal	41.5, 45.0, 7, 1576, .39, are terminating decimals	Decimal numbers Some are rational numbers, some are not rational numbers.
Repeating Decimals	.333, 0.51515151 = $0.\overline{51}$	a repeating decimal = rational number
Pi, Irrational Number	π = 3.14159 but does not repeat itself and is not a rational number. It is irrational.	Irrational numbers do not repeat.

the next 10 pages. After using this system for 10 pages, you may not need to label each page.

Follow these steps to improve your note-taking:

Step One – Record each problem step in the "Examples" section.

Step Two – Record the reasons for each step in the "Explanation/ Rules" section by using:

— abbreviations;

— short phrases, not sentences; and

— key words, properties, principles or formulas.

Step Three – Record key words/concepts in the left two-inch margin either during or immediately after lecture by reworking your notes.

Step Four – Cover up the "Example and Explanation" sections, and recite out loud the meaning of the key words or concepts.

Step Five – Place a check mark by the key words/concepts that you *did not* know.

Step Six – Review the information that you checked until it is under-stood.

Step Seven – Develop a Math glossary for difficult-to-remember key words and concepts.

A Math Glossary

The third area of focus in the seven steps to math note-taking is devoted to developing a math glossary. The math glossary is created to define a math vocabulary in your own words. Since math is considered a foreign language, understanding the math vocabulary becomes the key to comprehending math. Creating a glossary for each chapter of your textbook will help you understand math.

Your glossary should include all words printed in bold type in the text and any words you do not understand. If you cannot explain the math vocabulary in your own words, ask your instructor or tutor for help. You may want to use the last pages in your notebook to develop a math glossary for each chapter in your textbook. Review your math glossary before each test.

How to rework your notes

The note-taking system does not stop when you leave the classroom. As soon as possible after class, rework your notes. You can rework the notes between classes or as soon as you get home. By reworking your notes as soon as possible, you can decrease the amount of forgetting. This is an excellent procedure to transfer math information from short-term memory to long-term memory.

Remember: Most forgetting occurs right after learning the material. You need to rework the notes as soon as possible. Waiting means that you probably will not understand what was written.

The following are important steps in reworking your notes:

Step 1: *Rewrite the material you cannot read or will not be able to understand a few weeks later.* If you do not rework your notes, you will be frustrated when studying for a test if you come across notes you cannot read. Another benefit of rewriting the notes is that you immediately learn the new material. Waiting means it will take more time to learn the material.

Step 2: *Fill in the gaps.* Most of the time, when you are listening to the lecture, you cannot write down everything. It is almost impossible to write down everything even if you know shorthand. Locate the portions of your notes which are incomplete. Fill in the concepts which were left out. In the future, skip two or three lines in your notebook page for anticipated lecture gaps.

Step 3: *Add additional key words and ideas in the left-hand column.* These key words or ideas were the ones not recorded during the lecture.

> **Example:** You did not know you should add the *opposite* of 18 to solve a particular problem, and you incorrectly added 18. Put additional important key words and ideas (such as "opposite" and "negative of") in the notes; these are the words that will improve your understanding of math.

Step 4: *Add to your problem log those problems which the teacher worked in class.* The problem log is a separate section of your notebook that contains a listing of the problems (without explanations – just problems) which your teacher worked in class. If your teacher chose those problems to work in class, you can bet that they are considered important. The problems in this log can be used as a practice test for the next exam. Your regular class notes will not only contain the solutions but also all the steps involved in arriving at those solutions and can be used as a reference when you take your practice test.

Step 5: *Add calculator keystroke sequences to your calculator handbook.* The calculator handbook can be a spiral-bound set of note cards or a separate section of your notebook that holds only calculator-related information. Your handbook should also include an explanation of when that particular set of keystrokes is to be used.

Step 6: *Reflection and synthesis.* Once you have finished going over your notes, review the major points in your mind. Combine your new notes with your previous knowledge to have a better understanding of what you have learned today.

Use a Tape Recorder

If you have a math class during which you cannot get all the information down while listening to the lecture, ask your instructor about using a tape recorder. To ensure success, the tape recorder must have a tape counter and must be voice activated.

The tape counter displays a number indicating the amount of tape to which you have listened. When you find you are in an area of confusing information, write the beginning and ending tape counter number in the left margin of your notes. When reviewing your notes, the tape count number will be a reference point for obtaining information to work the problem. You can also reduce the time it takes to listen to the tape by using the pause button to stop the recording of unnecessary material.

Ask Questions

To obtain the most from a lecture, you must ask questions in class. By asking questions, you improve your understanding of the material and decrease your homework time. By *not* asking questions, you create for yourself unnecessary confusion during the remainder of the class period. Also, it is much easier to ask questions in class about potential homework problems than it is to spend hours trying to figure out the problems on your own at a later time.

If you are shy about asking questions in class, write down the questions and read them to your instructor. If the instructor seems confused about the questions, tell him/her you will discuss the problem after class. To encourage yourself to ask questions, remember:

— You have paid for the instructor's help;

— five other students probably have the same question;

— the instructor needs feedback on his/her teaching to help the class learn the material; and

— there is no such thing as a "stupid" question.

Record Each Problem Step

The final suggestion in note-taking is to record each step of every problem written or verbally explained. By recording each problem step, you begin *overlearning* how to work the problems. This will increase your problem-solving speed during future tests. If you get stuck on the homework, you will also have complete examples to review.

The major reason for recording every step of a problem is to understand how to do the problems while the instructor is explaining them instead of trying to remember unwritten steps. It may seem time consuming, however, it pays off during homework and test time.

How to correctly
use a calculator in class

In math, you are not only expected to learn the course material, you are also expected to learn how to use a calculator. This adds another dimension to note-taking in a math class that you will not encounter in humanities or other classes.

Do Not Stop Taking Notes

When a calculator is being used and demonstrated by an instructor while solving a problem, most students stop taking notes and attempt to duplicate the steps on their own calculators. *This is one of the biggest mistakes that a student can make.*

Unless you are very familiar with your calculator and know the location of important keys and functions, you will very likely get lost in the process of following along with the instructor. That old adage, "garbage in – garbage out" is universally true when using calculators. If you miss one keystroke, you might as well quit. You will not get the right answer.

Further, most students cannot take notes and manipulate the calculator at the same time. As a result, vast holes appear in your notebook where the explanations of the math involved and the interpretation of the answers for the problems should appear. And, finally, when you do your homework, you will probably find that you do not remember the exact sequence of keystrokes used to arrive at a particular answer.

Take Notes on Keystrokes

You can see how calculators can tremendously complicate the note-taking process. However, the solution to taking effective notes *and* becoming a proficient calculator user is quite simple. Instead of manipulating your calculator while the teacher is explaining, *write every keystroke down in your notes*.

Whenever possible, also write *why* a particular key was used to arrive at the answer and especially be sure to record the interpretation of the answer given by the calculator. (What does that number mean in the context of the problem being solved?)

Add Keystrokes to the Abbreviations List

You will certainly want to expand your own abbreviation list to include calculator keystrokes. In most cases, what is printed on the key can be used; however, you might consider using some of the following, especially if you are using a graphing calculator:

E for **[ENTER]**

G for **[GRAPH]**

AL for **[ALPHA]**

You can see that recording keystrokes, rather than executing them while the teacher is explaining, solves the note-taking problems listed earlier. You do not miss steps or get behind. You can continue to take notes the whole time, and when you do your homework, the step sequence is in your notes for your use.

Summary

— Effective listening is the first step to excellent note-taking.

— The effective listener knows where to sit in the classroom (The Golden Triangle of Success) and practices good listening techniques.

— The goal of note-taking is to write the least amount possible to record the most information.

— This allows you to enhance your ability to listen to the lecture and increase your learning potential in the classroom.

— Figure 11 (Modified Three-Column Note-Taking Sample), p. 121, is an excellent example to use for both taking notes and testing yourself on the information.

— Make sure you practice covering up the left side of the note page and recalling the information.

— The information you cannot recall must to be learned. Do not waste your time studying information you already know.

— Rework your notes as soon as possible after class.

— If you wait too long to review your notes, you might not understand them and it will be more difficult to learn them.

— When reworking your notes, make sure you complete your problem log; it will become very important when preparing for tests.

— Reworking your notes will improve not only your understanding of math but also your grades in the course.

— Calculator usage in the classroom has become a very effective learn-ing tool; however, it is difficult to take notes and use your calcula-tor at the same time.

— With complicated calculator usage, it is more important to write down the keystrokes rather than try to learn them in the classroom.

— Record these difficult calculator steps in your notes and rewrite them in your calculator log; then, practice the calculator steps until you know them.

Remember: The better note-taking skills you have, the more math you can learn in the classroom.

Assignment for Chapter 5

1. Review Figure 10 (Abbreviations), p. 118.

2. Review Figure 11 (Modified Three-Column Note-Taking Sample), p. 121.

3. How can you become an effective listener?

4. Why do you need to copy down each step of the math homework?

5. How does the use of a calculator complicate note-taking, and how can you solve this problem?

6. What abbreviations do you use in your math notes?

7. List and describe the seven steps to math note-taking.

8. What cues do math instructors give to indicate important test information?

9. How does asking questions in math class decrease the time you will spend on homework?

10. Who is the classmate with whom you can compare math notes?

11. Write down two problems from your problem log.

12. Write down an example from your calculator log.

Chapter 6

How to Improve Your Reading and Homework Techniques

As mentioned previously, reading a math textbook is more difficult than reading other textbooks. Math textbooks are written differently than your English or social science textbooks. Math textbooks are condensed material which take longer to read.

Mathematicians can reduce a page of writing to one paragraph, using math formulas and symbols. To make sure you understood that same information, an English instructor would take that original page of writing and expand it into two pages. Mathematicians pride themselves on how little they can write and still cover the concept. This is one reason why it may take you two to three times as long to read your math text as it would any other text.

Remember: Reading your math text will take longer than reading your other texts.

Math students are expected to know how to do their homework; however, most math students do not have a homework system. Most students begin their homework by going directly to the problems and trying to work them. When they get stuck, they usually quit. This is not a good homework system. A good homework system will improve your homework success and math learning at the same time.

Calculators are now required in most math classrooms. This is especially true of the graphing calculator. Students need to know how to use the calculator as a math tool when doing their homework and in testing situations. Knowing how to use your calculator during homework can improve your math learning; not knowing how to use a calculator on a test can cost you points.

In this chapter you will learn:

—   how to read a math textbook,

—   how to do your homework,

—   more ways to effectively use a calculator,

—   how to solve word problems,

—   how to recall what you have learned, and

—   how to work with a study buddy.

How to read
a math textbook

The way you read a math textbook is different from the traditional way students are taught to read textbooks in high school or college. Students are taught to read quickly or skim the material. If you do not understand a word, you are supposed to keep on reading.

Instructors of other courses want students to continue to read so they can pick up the unknown words and their meanings from context.

This reading technique may work with your other classes, but using it in your math course will be totally confusing. By skipping some major concept words or bold-print words, you will not understand the math textbook or be able to do the homework. Reading a math textbook takes more time and concentration than reading your other textbooks.

If you have a reading problem, it would be wise to take a developmental reading course before taking math. This is especially true with math reform delivery, where reading and writing are more emphasized.

Reform math classes deal more with word problems than do traditional math courses. If you cannot take the developmental reading course before taking math, then take it during the same semester as the math course.

Eight Steps to Understanding Reading materials

There are several appropriate steps in reading a math textbook:

Step 1 – *Skim the assigned reading material.* Skim the material to get the general idea about the major topics. Read the chapter introduction and each section summary. You do not want to learn the material at this time; you simply want to get an overview of the assignment. Then think about similar math topics that you already know.

> **Example:** Skimming will allow you to see if problems presented in one chapter section are further explained in the next chapter sections.

Step 2 – *As you skim the chapter, circle (using pencil) the new words that you do not understand.* If you do not understand these new words after reading the assignment, then ask the instructor for help. Skimming the reading assignments should take only five to 10 minutes.

Step 3 – *Put all your concentra-*
tion into reading. While reading the textbook, highlight the
material that is important to
you. However, do not highlight
more than 50 percent of a page
because the material is not
being narrowed down enough
for future study. Especially
highlight the material that is
also discussed in the lecture.
Material discussed both in the

textbook and lecture usually appears on the test. The purpose for
highlighting is to emphasize the important material for future study.
Do not skip reading assignments.

Remember: Reading a math textbook is very difficult. It might
take you half an hour to read and understand just *one* page.

Step 4 – *When you get to the examples, go through each step.* If the
example skips any steps, make sure you write down each one of
those skipped steps in the textbook for better understanding.
Later on, when you go back and review, the steps are already
filled in. You will understand how each step was completed. Also,
by filling in the extra steps, you are starting to overlearn the
material for better recall on future tests.

Step 5 – *Mark the concepts and words that you do not know.* Maybe
you marked them the first time while skimming. If you understand them now, erase the marks. If you do not understand the
words or concepts, then reread the page or look them up in the
glossary. Try not to read any further until you understand all the
words and concepts.

Step 6 – *If you do not clearly understand some words or concepts,*

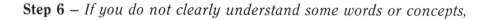

add these words to the note-taking glossary in the back of your notebook. Your glossary will contain the bold print words that you do not understand. If you have difficulty understanding the bold-print words, ask the instructor for a better explanation. You should know all the words and concepts in your notebook's glossary before taking the test.

Step 7 – *If you do not understand the material, follow these eight points, one after the other, until you do understand the material:*

Point 1 – Go back to the previous page and reread the information to maintain a train of thought.

Point 2 – Read ahead to the next page to discover if any additional information better explains the misunderstood material.

Point 3 – Locate and review any diagrams, examples or rules that explain the misunderstood material.

Point 4 – Read the misunderstood paragraph(s) several times aloud to better understand their meaning.

Point 5 – Refer to your math notes for a better explanation of the misunderstood material.

Point 6 – Refer to another math textbook, computer software program or video tape that expands the explanation of the misunderstood material.

Point 7 – Define exactly what you do not understand and call your study buddy for help.

Point 8 – Contact your math tutor or math instructor for help in understanding the material.

Step 8 — *Reflect on what you have read.* Combine what you already know with the new information that you just read. Think about how this new information enhances your math knowledge. Prepare questions for your instructor on the confusing information. Ask those questions at the next class meeting.

By using this reading technique, you have narrowed down the important material to be learned. You have skimmed the textbook to get an overview of the assignment. You have carefully read the material and highlighted the important parts. You then added to your note-taking glossary unknown words or concepts.

Remember: The highlighted material should be reviewed before doing the homework problems, and the glossary has to be learned 100 percent before taking the test.

How Reading Ahead Can Help

Reading ahead is another way to improve learning. If you read ahead, do not expect to understand everything. Read ahead two or three sections and put question marks (in pencil) by the material you do not understand.

When the instructor starts discussing that material, have your questions prepared and take good notes. Also if the lecture is about to end, ask the instructor to explain the confusing material in the textbook. Reading ahead will take more time and effort, but it will better prepare you for the lectures.

How to Establish Study-Period Goals

Before beginning your homework, establish goals for the study period. Do not just do the homework problems.

Ask yourself this question: "What am I going to do tonight to become more successful in math?"

By setting up short-term homework goals and reaching them, you will feel more confident about math. This also improves your self-esteem and helps you become a more internally motivated student. Set up homework tasks which you can complete. Be realistic.

Study-period goals are set up either on a time-line basis or an item-line basis. Studying on a time-line basis is studying math for a certain amount of time.

> **Example:** You may want to study math for an hour, then switch to another subject. You will study by time-line basis.

Studying by item-line basis means you will study your math until you have completed a certain number of homework problems.

> **Example:** You might set a goal to study math until you have completed all the odd problems in the chapter review. The odd problems are the most important problems to work. These, in most texts, are answered in the answer section in the back of the book. Such problems provide the opportunity to recheck your work if you do not get the answer correct. Once you have completed these problems, do the even-numbered problems.

No matter what homework system you use, remember this important rule: Always finish a homework session by understanding a concept or doing a homework problem correctly.

Do not end a homework session with a problem you cannot complete. You will lose confidence since all you will think about is the last problem you could not solve instead of the 50 problems you correctly solved. If you did quit on a problem you could not solve, return and rework problems you have done correctly.

Remember: Do not end your study period with a problem you could not complete.

How to Do
Your Homework

Doing your homework can be frustrating or rewarding. Most students jump right into their homework, become frustrated and stop studying. These students usually go directly to the math problems and start working them with out any preparation. When they get stuck on one problem, they flip to the back of the text for the answer. Then, they either try to work the problem backward, to understand the problem steps, or they just copy down the answer.

Other students go to the solution guide and just copy the steps. After getting stuck several times, these students will inevitably quit doing their homework assignment. Their homework becomes a frustrating experience, and they may even quit doing their math homework altogether.

To improve your homework success and learning, refer to the following 10 steps.

10 Steps to Doing Your Homework

Step 1 — *Review the textbook material that relates to the homework.* A proper review will increase the chances of successfully completing your homework. If you get stuck on a problem, you will have a better chance of remembering the location of similar problems. If you do not review prior to doing your homework, you could get stuck and not know where to find help in the textbook.

Remember: To be successful in learning the material and in completing homework assignments, you must first review your textbook.

Step 2 — *Review your lecture notes that relate to the homework.* If you could not understand the explanation of the textbook on how to complete the homework assignment, then review your notes.

Remember: Reviewing your notes will give you a better idea about how to complete your homework assignment.

Step 3 – *Do your homework as neatly as possible.* Doing your homework – neatly – has several benefits. When approaching your instructor about problems with your homework, he or she will be able to understand your previous attempts to solve the problem. The instructor will easily locate the mistakes and show you how to correct the steps without having to decipher your handwriting. Another benefit is that, when you review for midterm or final exams, you can quickly relearn the homework material without having to decipher your own writing.

Remember: Neatly prepared homework can help you now and in the future.

Step 4 – *When doing your homework, write down every step of the problem.* Even if you can do the step in your head, write it down anyway. This will increase the amount of homework time, but you are overlearning how to solve problems, which improves your memory. Doing every step is an easy way to memorize and understand the material. Another advantage is that ,when you rework the problems you did wrong, it is easy to review each step to find the mistake.

Remember: In the long run, doing every step of the homework will save you time and frustration.

Step 5 – *Understand the reasons for each problem step and check your answers.* Do not get into the bad habit of memorizing how to do problems without knowing the reasons for each step. Many students are smart enough to memorize procedures required to complete a set of homework problems. However, when similar homework problems are presented on a test, the student cannot solve the problems. To avoid this dilemma, keep reminding yourself about the rules, laws, or properties used to solve problems.

Example: *Problem:* 2 (a + 5) = 0. What property allows you to change the equation to 2a + 10 = 0? *Answer:* The distributive property.

Once you know the correct reason for going from one step to another in solving a math problem, you can answer any problem requiring that property. Students who simply memorize how to do problems instead of understanding the reasons for correctly working the steps will eventually fail their math course.

How to Check Your Answers

Checking your home work answers should be part of your homework duties. Checking your answers can improve your learning and help you prepare for tests.

Check the answers of the problems for which you do not have the solutions. This may be the even-numbered or odd-numbered problems or the problems not answered in the solutions manual.

First, check your answer by estimating the correct answer.

Example: If you are multiplying 2.234 by 5.102 the answer should by a little over 10. Remember that 2 times 5 is 10.

You can also check your answers by substituting the answer back into the equation or doing the opposite function required to answer the question. The more answers you check, the faster you will become. This is very important because increasing your answer checking speed can help you catch more careless errors on future tests.

Step 6 – *If you do not understand how to do a problem refer to the points on the next page.*

Point 1 — Review the textbook material that relates to the problem.

Point 2 — Review the lecture notes that relate to the problem.

Point 3 — Review any similar problems, diagrams, examples or rules that explain the misunderstood material.

Point 4 — Refer to another math textbook, solutions guide, math computer program software or video tape to obtain a better understanding of the material.

Point 5 — Call your study buddy.

Point 6 — Skip the problem and contact your tutor or math instructor as soon as possible for help.

Step 7 — *Always finish your homework by successfully completing problems.* Even if you get stuck, go back and successfully complete previous problems before quitting. You want to end your homework assignment with feelings of success.

Step 8 — *After finishing your homework assignment, recall to yourself or write down the most important learned concepts.* Recalling this information will increase your ability to learn these new concepts. Additional information about Step 8 will be discussed later in this chapter.

Step 9 — *Make up note cards containing hard-to-remember problems or concepts.* Note cards are an excellent way to review material for a test. More information on the use of note cards as learning tools is presented later in this chapter.

Step 10 — *Getting behind in math homework is academic suicide.* As mentioned in Chapter 1, math is a sequential learning process. If you get behind, it is difficult to catch up because each topic builds on the next. It would be like going to Spanish class without learning the last set of vocabulary words. The teacher would be talking to you using the new vocabulary, but you would not understand what was being said.

Do Not Fall Behind

To keep up with your homework, it is necessary to complete the homework every school day and even on weekends. Doing your homework one-half hour each day for two days in a row is better than one hour every other day.

If you have to get behind in one of your courses, *make sure it is not math*. Fall behind in a course that does not have a sequential learning process, such as psychology or history. After using the 10 Steps to Doing Your Homework, you may be able to combine two steps into one. Find your best combination of homework steps and use them.

Remember: Getting behind in math homework is the fastest way to fail the course.

More ways to effectively use a calculator

Most math classes require the use of graphing calculators. Calculators, if properly used, can improve student learning. The calculator can decrease careless errors and make it easy to check your answers. You can spend more time working problems and learning the concepts.

Most students, as pointed out in the last chapter, do not know how to use a calculator properly and waste valuable learning time trying. To help you better understand calculator usage, review the following important 10 steps before entering the classroom.

10 Steps to Using a Calculator

Step 1 – *Take your calculator and manual with you to class every day.* Even though on some class days you will not use the calculator, it is important to be ready to use the calculator to follow instructor demonstrations. More learning occurs when you can participate in problem-solving using your calculator.

Step 2 – *Resist the urge to individually use your calculator for "discovery" during class.* While personal calculator discovery is important, it should be done outside of class. Class time is best utilized for learning the essential algebraic processes for problem solving. Your other classroom participation should be in the form of listening, note-taking, questions and answers, or group work.

Step 3 – *When the instructor is demonstrating keystroke sequences, follow along on your calculator.* The best way to master the graphing calculator is by practicing – not by watching someone else do it. However, take notes on complicated keystroke sequences and practice them later.

Step 4 – *Keep pace with the instructor when working a problem using the graphing calculator.* You may be somewhat familiar with the keystroke sequence and find the instructor's pace a bit slow since he/she may be pacing the instruction for the benefit of students who are not at all familiar with the calculator. Be patient and do not jump ahead. Jumping ahead on a problem usually causes confusion.

Step 5 – *Be sure you understand key stroke sequences for your calculator if it is different from your instructor's.* It is not an efficient use of class time if the instructor is helping you with a particular keystroke while the other students have an altogether different calculator. Many instructors tell students at the beginning of the semester that they have expertise in only one calculator and cannot help a student who has a different type. You will need to rely on your calculator manual for needed help if this is the case.

Step 6 – *If you get lost using your calculator, ask the instructor for help.* Remember, when you register for the course, you paid for the instructor to help you when help is needed. Besides, there may be other students in the same predicament. On the other hand, use your best judgement in interrupting class for help. Sometimes exiting from your current screen and starting over can easily eliminate a class interruption.

Step 7 – *Class sessions that involve extensive calculator usage should be tape recorded.* It is virtually impossible to both take notes and follow along with your calculator during a demonstration of keystrokes. It is more important to participate with your calculator during class and replay the recording at home if you cannot write down the steps during instruction. Tape recording may not be necessary if your instructor distributes handouts of the keystrokes/processes covered during class.

Step 8 – *Check your solutions to problems using the RULE OF 3 – algebraically, numerically and graphically.* If you solve a problem algebraically, for example: $5x + 3 (x - 7) = 0$, check it by putting the left side of the equation on your calculator and tracing to the x-intercept. Or, if you solve a problem graphically, verify its solution by setting up and scrolling a table of functions values on your calculator.

Step 9 – *Memorize keystroke sequences for frequently used built-in functions.* Just as it is important to memorize the laws of exponents to work with polynomials, it is also important to memorize important calculator functions. Many calculator problems can be solved using built-in functions which eliminate the need to refer to your manual. This is timesaving and more productive, leaving you more time to understand the problem steps.

Step 10 – *Determine the feasibility of your solutions obtained using your calculator.* It is sometimes easy to press the key next to the key you wanted. For example, if you get "1528 feet" for the length of a table, rework the problem. Make sure your answer makes sense.

How to solve word problems

The most difficult homework assignment for most math students is working story/word problems. Solving word problems requires excellent reading comprehension and translating skills.

Students often have difficulty substituting English terms for algebraic symbols and equations. But once an equation is written, it is usually easily solved. To help you solve word problems follow these 10 steps:

Step 1 – *Read the problem three times.* Read the problem quickly the first time as a scanning procedure. As you are reading the problem the

second time, answer these three questions:

1. *What is the problem asking me?* (Usually at the end of the problem)

2. *What is the problem telling me that is useful?* (Cross out unneeded information).

3. *What is the problem implying?* (Usually something you have been told to remember).

Read the problem a third time to check that you fully understand its meaning.

Step 2 – *Draw a simple picture of the problem to make it more real to you* (e.g., a circle with an arrow can represent travel in any form – by train, by boat, by plane, by car, or by foot).

Step 3 – *Make a table of information and leave a blank space for the information you are not told.*

Step 4 – *Use as few unknowns in your table as possible.* If you can represent all the unknown information in terms of a single letter, do so! When using more than one unknown, use a letter that reminds you of that unknown. Then write down what your unknowns represent. This eliminates the problem of assigning the right answer to the wrong unknown. Remember you have to create as many separate equations as you have unknowns.

Step 5 – *Translate the English terms into an algebraic equation using the list of terms in Figure 12 (Translating English Terms into Algebraic Symbols), p. 150, and Figure 13 (Translating English Words into Algebraic Expressions), p. 151.* Remember the English terms are sometimes stated in a different order than the algebraic terms.

Step 6 – *Immediately retranslate the equation, as you now have it, back into English.* The translation will not sound like a normal English phrase, but the meaning should be the same as the original problem. If the meaning is not the same, the equation is incorrect and needs to be rewritten. Rewrite the equation until it means the same as the English phrase.

Step 7 – *Review the equation to see if it is similar to equations from your homework and if it makes sense.* Some formulas dealing with specific word problems may need to be rewritten. Distance problems, for example, may need to be written solving for each of the other variables in the formula. Distance = Rate x Time; therefore, Time = Distance/Rate, and Rate = Distance/Time. Usually, a distance problem will identify the specific variable to be solved.

Step 8 – *Solve the equation using the rules of algebra.*

Remember: Whatever is done to one side of the equation must be done to the other side of the equation. The unknown must end up on one side of the equation, by itself. If you have more than one unknown, then use the substitution or elimination method to solve the equations.

Step 9 – *Look at your answer to see if it makes common sense.*

Example: If tax was added to an item, it should cost more or if a discount was applied to an item, it should cost less. Is there more than one answer? Does your answer match the original question? Does your answer have the correct units?

Step 10 – *Put your answer back into the original equation to see if it is correct.* If one side of the equation equals the other side of the equation, then you have the correct answer. If you do not have the correct answer, go back to Step 5.

Figure 12

Translating English terms Into Algebraic symbols

Sum	+
Add	+
In addition	+
More than	+
Increased	+
In excess	+
Greater	+
Decreased by	-
Less than	-
Subtract	-
Difference	-
Diminished	-
Reduce	-
Remainder	-
Times as much	x
Percent of	x
Product	x
Interest on	x
Per	/
Divide	/
Quotient	/
Quantity	( )
Is	=
Was	=
Equal	=
Will be	=
Results	=
Greater than	>
Greater than or equal to	≥
Less than	<
Less than or equal to	≤

Figure 13

Translating English Words Into Algebraic Expressions

English Words	Algebraic Expressions
Ten more than x	$x + 10$
A number added to 5	$5 + x$
A number increased by 13	$x + 13$
5 less than 10	$10 - 5$
A number decreased by 7	$x - 7$
Difference between x and 3	$x - 3$
Difference between 3 and x	$3 - x$
Twice a number	$2x$
Ten percent of x	$.10x$
Ten times x	$10x$
Quotient of x and 3	$x/3$
Quotient of 3 and x	$3/x$
Five is three more than a number	$5 = x + 3$
The product of 2 times a number is 10	$2x = 10$
One half a number is 10	$x/2 = 10$
Five times the sum of x and 2	$5(x + 2)$
Seven is greater than x	$7 > x$
Five times the difference of a number and 4	$5(x - 4)$
Ten subtracted from 10 times a number is that number plus 5	$10x - 10 = x + 5$
The sum of 5x and 10 is equal to the product of x and 15	$5x + 10 = 15x$
The sum of two consecutive integers	$(x) + (x + 1)$
The sum of two consecutive even integers	$(x) + (x + 2)$
The sum of two consecutive odd integers	$(x) + (x + 2)$

The most difficult part of solving word problems is translating part of a sentence into algebraic symbols and then into algebraic expressions. Review Figure 12 (Translating English Terms into Algebraic Expressions), p. 150, and Figure 13 (Translating English Words into Algebraic Expressions), p. 151.

How to recall
what you have learned

After completing your homework problems, a good visual learning technique is to make note cards. Note cards are 3" x 5" index cards on which you place information that is difficult to learn or material you think will be on the test.

On the front of the note card write a math problem or information that you need to know. Color code the important information in red or blue. On the back of the note card write how to work the problem or give an explanation of important information.

> **Example:** If you are having difficulty remembering the rules for multiplying positive and negative numbers, you would write some examples on the front of the note card with the answers on the back.

Make note cards on important information you might forget. Every time you have five spare minutes, pull out your note cards and review them. You can glance at the front of the card, repeat to yourself the answer and check yourself with the back of the card. If you are correct and know the information on a card, do not put it back in the deck. Mix up the cards you do not know and pick another card to test yourself on the information. Keep doing this until there are no cards left that you do not know.

If you are an auditory learner, then use the tape recorder like the note cards. Record the important information just like you would on the front of the note card. Then leave a blank space on the recording. Record the answer. Play the tape back. When you hear the silence, put the tape on pause. Then say the answer out loud to yourself. Take the tape player off pause and see if you were correct. You can use this technique in the car while driving to college or work.

Review What You Have Learned

After finishing your homework, close the textbook and try to remember what you have learned. Ask yourself these questions, "What major concepts did I learn tonight?" or "What test questions might the instructor ask on this material?"

Recall for about three to four minutes the major points of the assignment, especially the areas you had difficulty understanding. Write down questions for the instructor or tutor. Since most forgetting occurs right after learning the material, this short review will help you retain the new material.

How to work with a study buddy

You need to have a study buddy when you miss class or when doing your homework. A study buddy is a friend or classmate who is taking the same course. You can find a study buddy by talking to your classmates or making friends in the math lab.

Try to find a study buddy who knows more about math than you do. Tell the class instructor that you are trying to find a study buddy and ask which students make the best grades. Meet with your study buddy several times a week to work on problems and to discuss math. If you miss class, get the notes from your study buddy so you will not get behind.

Call your study buddy when you get stuck on your homework. You can solve math problems over the phone. Do not sit for half an hour or an hour trying to work one problem; that will destroy your confidence, waste valuable time and possibly alienate your study buddy. Think how much you could have learned by trying the problem for 15 minutes and then calling your study buddy for help. Spend, at the maximum, 15 minutes on one problem before going on to the next problem or calling your study buddy.

Remember: A study buddy can improve your learning while helping you complete the homework assignment. Just do not overuse your study buddy or expect that person to do your homework for you.

Summary

— Reading a math text book is more difficult than reading texts for other courses.

— Students who learn how to correctly read a math text will be able to improve their math learning and understanding of the lecture material.

— Using the "Eight Steps to Understanding Reading Material," p. 135, is an excellent way to comprehend your math text.

— After using and understanding these steps, you may be able to customize your own reading steps to make reading easier and more efficient.

— Establishing study period goals is an excellent way to successfully manage homework time.

— Make sure to set a goal of either completing a given number of problems or working on problems for a set amount of time.

— Setting up short term goals and accomplishing them is one of the best ways to gain control over math.

— Make sure you finish every homework session by working problems you can do.

— Using the "10 Steps for Doing Your Homework," p. 140, is a way to do your homework and learn math at the same time.

— These steps will decrease the chances of your doing the homework one day and two days later forgetting how to work the problems.

— Follow these 10 steps until you are comfortable using them and until they become a part of your normal study routine.

— Then you may want to adjust the 10 steps to make your own efficient homework system.

— Now, and even more so in the future, students who are graphing-calculator literate will have a learning and testing advantage over other students.

— You must learn how to effectively use a graphing calculator or graphing/programmable calculator to be at least equal with your fellow students.

— Using the "10 Steps to Using a Calculator," p. 145, is a "step" in the right direction.

— Taking a calculator course or attending calculator workshops, along with understanding your calculator manual, will improve your calculator knowledge.

Remember: You need to know how to use your calculator efficiently prior to the test instead of learning how to use it during the test.

Assignment for Chapter 6

1. Review Figure 12 (Translating English terms into Algebraic symbols), p. 150, and Figure 13 (Translating English Words into Algebraic Expressions), p. 151.

2. How is reading a math textbook different from reading other textbooks?

3. List and describe the eight steps to reading a math textbook?

4. Which type of study period goal do you usually set?

5. What do you need to do before starting your math homework?

6. What are the reasons for writing down every problem step while doing the homework?

7. List and describe the 10 steps for doing your math homework?

8. List and describe the 10 steps for using a calculator.

9. How can note cards/tape recorder be used to improve math learning?

10. Describe how you solve word problems?

11. What should you do after finishing your math homework?

12. Who is your study buddy and why did you chose this particular person?

Chapter 7

How to Create a Positive Study Environment

A positive home and college study environment can improve your learning experiences. Traditional study environments include an on-campus study area (library) and an off-campus study area (your room). Study environments have been expanded to include the math lab/Learning Resource Center (LRC) and the classroom environment itself.

These new study environments require students to learn how to use resources in the math lab/LRC such as computer programs, CD-ROMs and assessment instruments. The learning environment in the classroom has also changed under the new math standards, which emphasize more collaborative learning.

Students now have to learn how to benefit from their collaborative classroom learning experiences and how to make the best use of the math lab or LRC. To maximize learning, students need to learn how to effectively use their new study environments and learning resources.

In this chapter you will learn:

— how to choose your best study environments,

— the best order in which to study subjects,

— the benefits of study breaks,

— the best way to use your math lab or Learning Resource Center, and

— how to enhance collaborative classroom learning.

How to choose your
best study environments

Your study environment is the key to efficient learning. Some study environments have too many distractions, which cause lack of concentration and lost study time. Other study environments may be quiet, but you may not have the necessary materials around you to supplement your studies.

Creating a positive study environment may take some time, but it will improve the quality of studying. Choosing an appropriate study place may seem trivial, but it can significantly enhance learning. When you start studying, choose a specific place at home, at college and at the math lab or Learning Resource Center (LRC).

Choosing a Place to Study

While studying in your home, choose one place, one chair, one desk or table as your study area. If you use the kitchen table, choose one chair, preferably one that you do not use during dinner. Call this chair "my study chair." If you study in the student cafeteria, use the same table each time. Do not use the table at which you play cards or eat.

By studying at the same place each time, a conditioned response will be formed. From then on, when you sit down at your study place your mind will automatically start thinking about studying. This conditioned response decreases your "warm up" time. "Warm up" time is how long it takes to actually begin studying after you sit down.

Another aspect of the study environment involves the degree of silence you need for studying. In most cases, a totally quiet room is not necessary. But if you can study only with total silence, keep this in mind when selecting your study places and times.

Most students can study with a little noise, especially if it is a constant sound, like a mellow radio station. In fact, some students keep on a mellow radio station or a fan to drown out other noises. However, do not turn on the television to drown out other noises or to listen to while studying. That will not work! For most efficient studying, select a study area where *you* can control the noise level.

Setting Up Your Study Area

Your home study environment should be surrounded by signs that "tell" you to study. One sign should be your study schedule. Attach your study schedule to the inside flap of your notebook and place another copy in your study place at home. Place your study goals and the rewards for achieving those goals where they can be easily seen.

Do not post pictures of your girl friend, boy friend, bowling trophies, fishing trophies or other items in your study area; these could be distracting. Instead, post pictures indicating your goals after gradu-

ation. If you want to be a nurse, doctor, lawyer or business person, post pictures that represent these goals. The study area should always reinforce your educational goals.

When sitting down to study, have ready the "tools of your trade": pencils, paper, notebook, textbook, study guide and calculator. Anything you might need should be within reach. In this way, when you need something, you can reach for it instead of leaving your study area.

The problem with getting up is not just the time it takes to get the item, but the time it takes to "warm up" again. After getting milk and cookies and sitting down, it takes another four to five minutes to "warm up" and continue studying.

The best order in which to study subjects

When studying, arrange your subjects in the order of difficulty. In other words, start with your most difficult subject – which is usually math – and work toward your easiest course. By studying your most difficult subject first, you are more alert and better motivated to complete the work before continuing on the easier courses, which may be more interesting to you. If you study math last, you will probably tire easily, become frustrated and you may quit. However, you are less likely to quit when you study a subject that interests you.

Remember: Study math first!

Mix the Order of Study

Another approach to improving the quality of your study is to mix up the order of studying different subjects.

Example: If you have English, accounting and math to study, then study them in the following order: 1] math, 2] English, and 3] accounting. By studying the subjects in this order, one part of your brain can rest after studying math while the other part of your brain is studying English. Now your mind is "fresh" when you study accounting.

Decide When to Study

Deciding when to study different types of material is part of developing a positive study environment. Your study material can be divided into two separate types. One type is new material and the other type is material that has already been learned.

The best time to review the material you have already learned is right before going to sleep. By reviewing it the night before, you will have less brain activity and fewer physical detractors that would prevent you from recalling the material the next day.

Example: If you have an 8:00 test the next morning, you should review the material the night before. If you have a 10:00 test the next day, review the material both the night before and the day of the test. Reviewing is defined as reading the material to yourself. You also might review a few problems you have already solved to keep your mind alert, but do not try to learn any new material the night before the test.

When to Learn New Material

Learning new material should be done during the first part of the study period. Do not learn new material the night before a test. You will be setting yourself up for test anxiety. If you try to cram the procedures to solve different types of equations or new ways to factor trinomials, you will end up in a state of confusion. This is especially true if you have major problems learning the new equations or factoring. The next day you will only remember *not* being able to solve the equation or factor the trinomials; this could distract you on the test.

Most students get tired after studying for several hours or before going to bed. If you are tired and try to study new material, it becomes more difficult to retain. It takes more effort to learn new material when you are tired than it does to review old material. When you start getting tired of studying, the best tactic is to begin reviewing previously learned material.

Find the Most Efficient Time to Study Math

The most efficient time to study is as soon as possible after the math class. Psychologists indicate that most forgetting occurs right after learning the material. In other words, you are going to forget most of what you have learned in the first hour after class. To prevent this mass exodus of knowledge, you need to recall some of the lecture material as soon after class as is practical.

The easiest way to recall the lecture is to rework your notes. Reviewing your notes will increase your ability to recall the information and make it easier to understand the homework assignments.

Choosing Between Mass and Distributive Learning

There are two different types of learning processes: "mass learning" and "distributive learning." Mass learning involves learning everything at one time. Distributive learning is studying the same amount of time as mass earning — with study breaks.

> **Examples:** *Mass learning* – you would study three hours in a row without taking a break, then quit studying for the night. *Distributive learning* – you would study for about 50 minutes with a 10 minute break, study for 50 more minutes with a 10 minute break, and finish with 60 minutes of studying before stopping.

The benefits of study breaks

Psychologists have discovered that learning decreases if you do not take study breaks. Therefore, use the distributive learning procedures (described above) to study math.

If you have studied for only 15 or 20 minutes and feel you are not retaining the information or your mind is wandering, take a break. If you continue to force yourself to study, you will not learn the material. After taking a break, return to studying.

If you still cannot study after taking a break, review your purpose for studying and your educational goals. Think about what is required to graduate. It will probably come down to the fact that you will have to pass math. Think about how studying math today will help you pass the next test; this will increase your chances of passing the course and of graduating.

Write on an index card three positive statements about yourself and three positive statements about studying. Look at this index card every time you have a study problem. Use every opportunity available to reinforce your study habits.

The best way to use your math lab or Learning Resource Center

Learning how to use your math lab or Learning Resource Center (LRC) can improve your learning and, in turn, your grades. Many students are unaware of the tutorial and learning resources offered at their college or university. Some students find out about these resources after they are failing, which, in most cases, is too late. You need to find the location of learning resources and how to utilize them as soon as possible after course registration.

Some colleges and universities have math labs, LRCs, Academics Enrichment Centers, computer labs, Student Support Services, Disabled Student Services or other specialized labs to help students. You should ask your instructor and counselor where to get help in math.

Do not forget to ask your fellow students for recommendations to get help in math. Sometimes the students know the best places to get help.

In most colleges and universities there will be more than one place to get additional help in math. Visit all the places you hear about, each at various times so you can see how they operate with different personnel and under crowded or sparse conditions. Ask more than one person at each location what materials they have to help you. Find out if they have:

1. *Videos* – available from the textbook publisher, commercially bought or produced by local instructors.

2. *Computer Programs/CD-ROMs* – available either from the textbook publisher or commercially bought.

3. *Supplemental Texts.*

4. *Old Textbooks* on your course.

5. *Tutors, Lab Aides or Instructors* available.

6. *Practice Tests.*

7. *Assessments Instruments.*

8. *Other Helpful Items* – manipulatives, models, posters, graphing calculators or something new.

Students will find that some of the learning resources will be more effective than others. Use your learning modality and cognitive style preference to decide which learning resources to try first.

If you are a visual learner, then you should try the video tapes first. However, since you may still learn information through another learning style, you need to try all the learning resources to find out which ones work best.

The resources you select may be based more on their quality than on your learning style. Become familiar with all the resources and then make a decision on which resources best help you. You will find that some resources enhance your learning while other resources hinder your learning. To help you evaluate the resources, use Figure 14 (Math lab/LRC Check Sheet) on the next page.

Suggestions for using the above-listed eight resources:

1. *Videos* – Locate the video tapes which correspond to the section of the text you are learning. You may have to switch to different video tape resources based on the section of the text you are studying. Have the book open to the section of the text to be discussed on the video, and have paper and pencil ready to take notes. Use the fast forward button to bypass what you have mastered. When you reach the section that gives you difficulty, slow down and watch that section several times.

Figure 14
Math lab/LRC check sheet

Part One

Name of Lab_____ Phone _____

Location_____ Hours_____

Tutor/Student ratio_____Trained Tutors: Yes _____ No _____

Best time to attend _____ Worst time to attend _____

Required lab time _____

Check out materials: Video tapes _____ Computer programs _____

                         Other _____

Part Two

Check the available resources and rate the usefulness of each:

1. Video Tapes: Text _____ Commercial _____ Teacher_____ None _____
    (Rating: Poor 1 2 3 4 5 Excellent)

2. Tutorial Computer Programs:
    Text___Commercial___Teacher___None___
    (Rating: Poor 1 2 3 4 5 Excellent)

3. CD-ROM/Video Disk: Text___Commercial_____Teacher_____None_____
    (Rating: Poor 1 2 3 4 5 Excellent)

4. Old Textbooks: Yes _____ No_____
    (Rating: Poor 1 2 3 4 5 Excellent)

5. Manipulatives/Models: Text ___ Commercial ___ Teacher___
                           None___
    (Rating: Poor 1 2 3 4 5 Excellent)

Figure 14
Math lab/LRC check sheet, *continued*

6. Assessments: Placement Tests_____ Diagnostic Math Tests____
    Math Study Skills Evaluation _____ Learning Styles Inventories ____
    Math Anxiety Rating Scales _____
    (Overall Rating: Poor 1 2 3 4 5 Excellent)

7. Tutors: Student _____ Graduate Student____ Professional____
    (Rating: Poor 1 2 3 4 5 Excellent)

8. Overall Math Lab/LRC Experience: _____
    (Rating: Poor 1 2 3 4 5 Excellent)

Part Three

Describe the resource that you found most helpful:

_____

_____

_____

Remember: No one cares about the number of times you rewind the video tape.

If the topic still confuses you, note that topic and look for another resource. You might want to view another video tape from a lower-level text that explains the basics of the concept you are trying to learn. Also, ask for a local instructor's video tape or a video from a previous text that might explain that concept in words that you understand. If the video tapes do not help you, go to another resource.

2. *Computer Programs/CD-ROMs* – Locate the computer program that best goes with your text. It may be the computer software offered

by the textbook publisher or some commercially bought software like *Algeblaster III* (Davison and Associates, 1993). Ask if you can copy it and use it at another location or on your own computer.

Review the other available software programs to find the one that fits your needs. The newer computer programs are more user friendly, but some of the older programs are more effective.

Make sure your computer program gives you the reasons for missing an answer, not just the answer. Just as people have different personalities, computer programs offer different teaching methods. Choose the one you like best. Ask for any CD-ROM programs because they can provide both video and auditory learning.

3. *Supplemental Texts* – Most textbooks offer supplemental texts. The supplemental text may be a study guide or solutions manual. Review the supplemental text for intermediate steps needed to solve the homework problems. Make sure it explains how each step was obtained.

   You can check your homework, step-by-step, to find your errors. If you notice any step that you often miss, write it down and ask yourself if there is some concept involved that you do not understand. Talk to your instructor or tutor if the problem persists.

   Supplemental texts can be very helpful if you use them to understand how to do the problems instead of just another source for copying the steps and the answer.

4. *Old Textbooks* – Math is a universal language, but different textbooks describe the same topic with different English. You may understand another text better than your current text. Some textbook authors are often better at explaining a topic than are others. This is especially true if you are bilingual or raised in a different part of the country. If you can locate a math textbook on the same level and describing the same topic, it may be very helpful.

5. *Tutoring* – Most students prefer tutoring as their best learning resource. However, research has shown that untrained tutors are

no better than no tutors at all. Try to work with a trained tutor who has had your course. Explain to the tutor your learning style and suggest that he/she tutor you based on your learning style.

Example: If you are an auditory learner, then have the tutor orally explain to you how to solve the problem. Then repeat back what the tutor said (in your own words). Make sure the tutor does not just work the problem for you without explaining the reasons for each step.

Try not to schedule your tutoring sessions around lunch time since it is usually the busiest time of all.

Have your questions ready from your previous homework assignments. Focus on the concepts you do not understand, not just on how to work the problem. The more specific your are about your homework problems, the more tutorial help you will receive.

Do not expect miracles! If you tell your tutor, "I have a test in twenty minutes and do not understand anything about Chapter 6!" – about all the tutor can do is offer to pray for you. However, past experiences have shown that those who have previously helped themselves to tutoring are most likely to be rewarded with good grades.

Remember: Make sure you get help early. Waiting until you are failing a course to get help may be too late.

6. *Practice Tests* – Use practice tests to find out what you do not know before the real test. Ask if the math lab/LRC offers practice tests. Take these practice tests at least two days before the real test. This will give you at least one day to find out how to work the missed problems and another to review for the test.

The more realistic practice tests you can take, the better you will do on the real test. Make sure the practice tests are timed and do not use any of your notes or text.

Real testing conditions prepare you for the test just like the government prepares its soldiers for combat. The government spends a fortune sending its soldiers through mock combat because, by doing so, fewer casualties occur in real combat. Make sure you are not a testing casualty.

7. *Assessment Instruments* – Assessment instruments can be used to place you into the correct course, locate your math weakness and help you understand your learning strengths and weaknesses. If you are not sure that you have been placed into the correct course, ask to take a placement test. Being placed into the correct course is *a must* to pass math. Ask if the lab has diagnostic math tests to locate your weaknesses. Ask about other assessment instruments which can be used to help improve your learning.

8. *Other Helpful Items – Manipulatives, Models, Posters, Graphing Calculators.* If a picture is worth a thousands words, a model is worth a million for the tactile/concrete learner. Ask for what may be called "manipulatives" or "3-D models." Manipulatives and models are concrete representations (which you can physically touch) of a concept. Look for posters for additional information. Further, ask for connections for uploading graphing calculator programs from a computer. Ask if there is a calculator-based lab that can easily be used to explore physical applications of math in real time.

In general, learning math is a lot like learning to ride a bicycle. You can watch someone else do it, but you only learn by trying it yourself. You must believe in yourself and keep at it. Even if you are very wobbly at first, as long as you keep peddling, you are riding. But if you do not believe in yourself enough to keep peddling, you will fall. In time, you will be steady enough to take off the training wheels and wonder why you ever thought they were necessary. Even if you get rusty after a long absence, you will never again need training wheels.

Math is also something you learn by trying it yourself. Others can assist you with techniques, but you need to learn to *solo*. As long as you keep trying, you are learning to think mathematically, and you will be able to do it. In the future, even after a long time away from math, you will remember that you were able to master math before, and, with a little review, you still can.

How to enhance collaborative classroom learning

Collaborative learning is a mode of learning that involves student participation in a small group to complete a desired task, assignment or project. In a math class, collaborative learning would be a small group of two to six students working together to solve math problems. Due to math reforms, math classes will include more collaborative learning as a mode of instruction.

Example 1: You develop a study group that is preparing for an upcoming test. Each group member makes up several sample test questions on note cards and the group discusses the answers.

Example 2: You are in the math classroom and the class is divided into groups of four. Each group is assigned a different word problem to solve. The student recorder listens to the group discussing the problem and writes down the solution steps. Each group shares the solution with the other groups.

There are different types of collaborative learning exercises which your instructor may use. Each exercise involves two or more students working together to solve some type of problem. Your instructor may combine traditional instruction and collaborative learning exercises. Collaborative learning has some benefits over traditional instruction.

The Benefits of Collaborative Learning

— Less fear of asking questions in a small group compared to asking questions in the classroom,

— You may have a group learning style that enhances your learning.

— Sometimes an explanation of a concept or problem is more effective coming from a group member than from the instructor.

— Your instructor is free to walk around and individually help group members with difficult questions.

— The group can make up test questions and quiz each other.

— Learning groups can lead to after-class study groups.

— Collaborative learning prepares you for the business and industry work force. Experience with groups and team-building is important to a prospective employer.

Being a good collaborative learner can be different from being a good individual learner. To benefit from collaborative learning, you may have to learn some new skills.

Characteristics of Good Group Members

— Completing any necessary preparation work prior to your group meeting. Little is accomplished in a group meeting if individual commitments are broken.

— Being supportive and acknowledging participation of fellow group members' ideas even if they are different from yours.

– Encouraging the group to stay on task. If discussion strays, lead the group back toward your team goal.

– Keeping a good balance between being an active participant and a good listener. Both of these characteristics are critical to a positive collaborative learning environment. Speak for yourself and let the others speak for themselves.

– Accepting help and suggestions from other members without feeling guilty.

– Accepting group members who try even if they cannot solve the problems.

– Bringing closure to a team session by summarizing the group's efforts. Reach consensus (everyone agrees) on any group decisions involving completion of your tasks.

Since this is a new learning experience, some group members may not participate or may become disruptive. If a group member on your team is not participating or is being disruptive, follow the five steps below, in order.

Five Steps for Handling a Disruptive Group Member

1. Discuss the matter out in the open as a group. Remember, problems do not get solved by ignoring them or disguising them.

2. Ask the instructor to review the his/her own collaborative learning rules or the seven behaviors of good group members, emphasizing how only one group member could be a detriment to the learning process.

3. Ask the instructor to attend a conference with the group con-

cerning the disruptive member. Invite the disruptive member to attend the conference. Each person can share their views on the issue and establish possible solutions.

4. If the disruptive group member does not attend the conference, ask the instructor to individually talk with the group member.

5. If the problem still persists, ask the instructor to remove the disruptive member from the group.

Some students who are individual learners may have difficulty working in groups. This situation is similar to that of a student who is a visual learner being in a classroom where the instructor mostly lectures.

Even if you are an individual learner, you can still benefit from collaborative learning. One way is to volunteer for tasks that require individual work. For example, if your group's task is to develop sample test questions, you might ask to develop yours alone or provide the answers to all the questions. Another way to be an active member is to be the group recorder. This is an individual task that is still a contribution to the group. A third way is to take an active role in teaching other students how to solve problems.

Remember: A student who tutors another student on solving a math problem will seldom miss that problem on the test.

Finally, if you are having difficultly learning the material, you can still go to the instructor, tutor or math/LRC for individual help. However, since team work is an important skill in business/industry and other careers, it is vital that you practice these skills.

Summary

- A positive study environment can improve your math grades.

- Establishing several appropriate study places can increase your learning potential.

- Using distributive learning and studying new and old material at the appropriate times can improve your learning skills.

- Studying at the most efficient time, which is right after math class, can improve your learning.

- Using the new study environments to your advantage can increase your knowledge.

- Due to math reforms, collaborative learning is going to become an important part of the math classroom environment.

- Being an effective collaborative learner not only will help you in math, it will help you in your future career.

- Learning how to use the math lab/LRC resources which match your learning modality enhances learning. Therefore, make sure you try each resource at least once to see how it helps you.

- Study skills and math lab/LRC resources can compensate for a mismatch of teaching and learning styles.

- For additional information, read Reference A – "10 Steps to Improving Your Study Skills," p. 281, and Reference B – "Suggestions to Students for Improving Math Study Skills," p. 291.

Remember: The more you use the appropriate math lab/ LRC resources, the better math grades you will make.

Assignment for Chapter 7

1. Read Reference A – "10 Steps to Improving Your Study Skills," p. 281.

2. Read Reference B – "Suggestions to Students for Improving Math study skills," p. 291.

3. Complete Figure 14 (Math lab/LRC Check Sheet), p. 168-69.

4. How can you improve your study environment?

5. In what order do you need to study your courses?

6. When should you learn new material?

7. When is the most efficient time to study math?

8. What should you do when you cannot study?

9. What are your best math lab/LRC resources?

10. What are the seven characteristics of good group members?

Chapter 8

How to Remember What You Learn

To understand the learning process, you must understand how your memory works. You learn by conditioning and thinking. But memorization is different from learning. For memorization, your brain must perform several tasks, including receiving the information, storing the information and recalling the information.

By understanding how your memory works, you will be better able to learn at which point your memory is failing you. Most students usually experience memory trouble between the time their brain receives the information and the time it is stored.

There are many techniques for learning information that can help you receive and store information without losing it in the process. Some of these techniques may be more successful than others, based on you and how you best learn.

In this chapter you will discover:

— how you learn,

— how short-term memory affects what you remember,

— how working memory affects what you remember,

— how long-term memory/reasoning affects what you remember,

— how to use memory techniques,

— how to develop practice tests, and

— how to use number sense.

How you learn

Educators tell us that learning is the process of "achieving competency." More simply put, it is how you become good at something. The three ways of learning are by conditioning, by thinking and by a combination of conditioning and thinking.

Learning by Conditioning and Thinking

Conditioning is learning things with a maximum of physical and emotional reaction and a minimum of thinking.

Examples: Repeating the word "pi" to yourself and practicing where the symbol is found on a calculator are two forms of conditioned learning. You are learning using your voice and your eye-hand coordination (physical activities), and you are doing very little thinking.

Thinking is defined as learning with a maximum of thought and a minimum of emotional and physical reaction.

> **Example:** Learning about "pi" by thinking is different than learning about it by conditioning. To learn "pi" by thinking, you would have to do the calculations necessary to result in the numeric value which the word "pi" represents. You are learning *using your mind* (thought activities), and you are using very little emotional or physical energy to learn "pi" in this way.

The most successful way to learn is to combine thinking and conditioning. The best learning combination is to learn by thinking first and conditioning second.

Learning by thinking means that you learn by

— *observing*,

— *processing*, and

— *understanding* the information.

How Your Memory Works

Memory is different from learning; it requires reception, storage and retrieval of information. The way you *receive* information is accomplished through your five senses (known as the *sensory input*): what you see, feel, hear, smell and taste.

> **Examples:** In math classes, you will use your sense of *vision* to both watch the instructor demonstrate problems on the chalkboard and to read printed materials. You will use your sense of *hearing* to listen to the instructor and other students discuss the problems. Your sense of *touch* will be used to operate your calculator and to appreciate geometric shapes. In chemistry and other classes, however, you may additionally use your senses of *smell* and *taste* to identify substances.

The sensory register briefly holds an exact image or sound of each sensory experience until it can be processed. If the information is not processed immediately, it is forgotten. The sensory register helps us go from one situation to the next without cluttering up our minds with trivial information.

Processing the information involves placing it in either short-term memory, working memory or long-term memory.

How short-term memory affects what you remember

Information that passes through the sensory register is stored in short-term memory. Remembering something for a short time is not hard to do for most students. By conscious effort, you can remember math laws, facts and formulas received by the sensory register (your five senses) and put them in short-term memory. You can recognize them and register them in your mind as something to remember for a short time.

> **Example:** When you are studying math, you can tell yourself the distributive property is illustrated by a(b+c)=ab+ac. By deliberately telling yourself to remember that fact (by using conditioning – repeating or writing it again and again), you can remember it, at least for a while, because you have put it in short-term memory.

Psychologists have found that short-term memory cannot hold an unlimited amount of information. You may be able to use short-term memory to remember one phone number or a few formulas but not five phone numbers or 10 formulas.

Items placed into short-term memory usually fade fast, as the name suggests.

Examples: Looking up a telephone number in the directory, remembering it long enough to dial, then forgetting it immediately. Learning the name of a person at a large party or in a class but forgetting it completely within a few seconds. Cramming for a test and forgetting most of it before taking the test.

Short-term memory involves the ability to recall information immediately after it is given (without any interruptions). It is useful in helping you concentrate on a few concepts at a time, but it is not the best way to learn. This is especially true of students who have attention problems or problems with short-term memory.

Students with short-term memory problems forget a math step as soon as the instructor explains it. To remember more facts or ideas, and keep them in memory a longer period of time – especially at test time – use of a better system than short-term memory is required.

Changes in the way math is taught include the idea that conditioning learning (the most often used teaching form) leads only to short-term memory. Conditioning learning is often used for short-term recall of concept skills and can only be stored for longer periods of time by thinking. Thinking takes place when applying skills to working the problems.

How working memory affects what you remember

Working memory (or long-term retrieval) is that part of the brain that can work on problems for a longer period of time than can short-term memory. Working memory, then, offers an increase in the amount of *time* information is held in memory. (An increase in the *volume* of information that can be held requires long-term memory.)

Working memory involves the ability to recall information after learning has been consistently interrupted over a period of several minutes. Students with working memory problems may listen to a math lecture and understand each step as it is explained. When the instructor goes back to a previous step discussed several minutes prior, however, the student has difficulty explaining or remembering the reasons for the steps. These students have difficulty remembering series steps long enough to understand the concept.

**Working memory:
Your internal chalkboard**

Working memory can be compared to a mental work space or an internal chalkboard. Just like a chalkboard, working memory has limited space, which can cause a "bottleneck" in learning.

Working memory uses the information (such as multiplication tables) recalled from long-term memory. Overlearning math facts – by storing them in long-term memory – can free up more working memory to solve problems.

Example: In solving 26 x 32, you would put the intermediate products 52 (*from 2 x 26*) and 780 (*from 30 x 26* – remember 3 is in the 10's place, so make it 30) into working memory and add them together. The more automatic the multiplication, the less working memory you use. If you cannot remember your multiplication, you use up working memory trying to solve the multiplication problem. This leaves you with less working memory to solve the resulting addition problem.

How long-term memory/reasoning affects what you remember

Long-term memory is a storehouse of material that is retained for long periods of time. Long-term memory is recalled into working memory to solve problems. It is *not* a matter of trying harder and harder to remember more and more unrelated facts or ideas; it is a matter of organizing your short-term memories and working memories into meaningful information.

Reasoning is thinking about memories, comprehending their meanings and understanding their concepts. Without rehearsing the information, it will not be processed into long-term memory/ reasoning.

The main problem students face is converting learned material from short-term memory or working memory into long-term memory — and understanding it. To place information into long-term memory, you need to understand it and effectively concentrate on it.

Converting material from short-term memory to long-term memory

Remember: Securing math information into long-term memory is not accomplished by just doing the homework — you must also understand it.

The Role of "Memory Output" in Testing

Memory output is what educators call a "retrieving process." It is necessary for verbal or written examinations. It is the method by which you put information stored in your long-term memory onto paper or into words. Memory output can be blocked by three things:

1. insufficient processing of information into long-term memory,

2. test anxiety, or

3. poor test-taking skills.

If you did not completely place all of the information you learned into long-term memory, you may not be able to give complete answers on tests.

Test anxiety can decrease your ability to recall important information or it can cause you to block out information totally. Ways to decrease test anxiety will be discussed in Chapter 9, "How to Reduce Math Test Anxiety." Students who work on their test-taking skills can improve their memory output. Techniques to improve both your memory output and test-taking skills will be explained in Chapter 10, "How to Improve Your Math Test-Taking Skills."

Understanding the Stages of Memory

Understanding the stages of memory will help you answer this common question about learning math: "How can I understand the procedures to solve a math problem one day and forget how to solve a similar problem two days later?"

There are three good answers to this question. First, after initially learning how to solve the problem, you did not *rehearse* the solving process enough for it to enter your long-term memory. Second, you did get the information into long-term memory, but the infor-

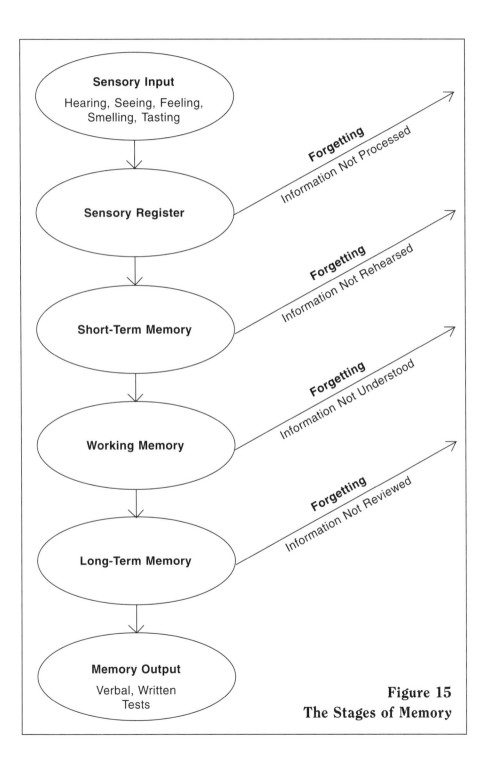

Figure 15
The Stages of Memory

mation was not *reviewed* frequently enough and was, therefore, forgotten. Third, you memorized how to work the problem but did not *understand* the concept. See Figure 15 (The Stages of Memory), p. 189.

How to use
memory techniques

There are many different techniques which can help you store information in your long-term memory. Having a positive attitude about studying, using your best learning sense(s), decreasing distractions and other techniques can improve your long-term memory.

A Good Study/Math Attitude

Having a positive attitude about studying will help you concentrate and improve your retention. This means you need to have at least a neutral math attitude (you neither like nor dislike it), and you should reserve the right to actually learn to *like* math. View studying as an opportunity to learn rather than as an unpleasant task. Tell yourself that you *can* learn the material and that *learning it will help you pass the course and graduate.*

Using Your Best "Learning Sense"

Using your best *learning sense* (what educators call your "predominate learning modality") can improve how well you learn and enhance the transfer of knowledge into long-term memory/reasoning. Your learning senses are: Vision, hearing, touching, etc. Ask yourself if you learn best by watching (vision), listening (hearing), or touching (feeling).

Visual (watching) Learner

Knowing that you are a visual math learner can help you select the memory technique that will work best for you. Repeatedly reading and writing math materials being studied is the best way for a visual learner to study.

A visual way to decrease distractions is using the "my mind is full" concept. Imagine that your mind is completely filled with thoughts of learning math, and other distracting thoughts cannot enter. Your mind has one-way input and output, which only responds to thinking about math when you are doing homework or studying.

Auditory (hearing) Learner

If you are an *auditory learner* (one who learns best by hearing the information), then learning formulas may be best accomplished by repeating them back to yourself or recording them on a tape recorder and listening to them. Reading out loud is one of the best auditory ways to get important information into long-term memory. Saying facts and ideas out loud improves your ability to think and remember. If you cannot recite out loud, recite the material to yourself, emphasizing the key words.

An auditory way to improve your concentration is by becoming aware of your distractions and telling yourself, out loud, to concentrate. If you are in a location where talking out loud will cause a disturbance, mouth the words "start concentrating" as you say them in your mind. Your concentration periods should increase.

Tactile/concrete (touching) Learner

A *tactile/concrete learner* needs to feel and touch the material to learn it. Unfortunately, this learning sense is not used by most

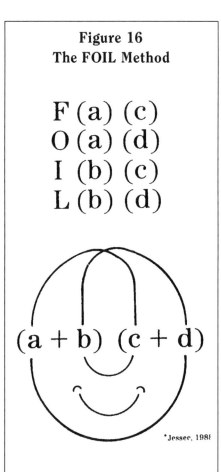

Figure 16
The FOIL Method

F (a) (c)
O (a) (d)
I (b) (c)
L (b) (d)

(a + b) (c + d)

*Jessee, 198!

FOIL is used to remember the procedure to *multiply to binomials*. The letters in FOIL stand for **F**irst, **O**utside, **I**nside, and **L**ast. To use FOIL, multiply the following:

— the **F**irst terms ((a) (c)),
— the **O**utside terms ((a) (d)),
— the **I**nside terms ((b) (c)),
— the **L**ast terms ((b) (d)).

To learn FOIL, trace your finger along the FOIL route.

math instructors. As a result, students who depend heavily upon feeling and touching for learning will have the most difficulty developing effective learning techniques. For example, if you want to learn the FOIL method (see Use Acronyms, p. 195), you would take your fingers and trace the "face" to remember the steps. See Figure 16 (The FOIL Method).

A tactile/concrete way to improve your study concentration is by counting the number of distractions for each study session. Place a sheet of paper by your book when doing homework. When you catch yourself not concentrating, put the letter "C" on the sheet of paper. This will remind you to concentrate and get back to work. After each study period, count up the number of "C's" and watch the number decrease.

Using Multiple Senses

If you have difficulty learning material from one sense, you might want to try learning materials through two or three senses. Involving two or more senses in learning improves your learning and remembering. Review Reference C – "Student Learning-Style

Information," p. 295, for math-learning suggestions based on your senses.

If your primary sense is visual and your secondary sense is auditory, you may want to write down equations while saying them out loud to yourself. Writing and reciting the material at the same time combines visual and auditory and some tactile/concrete styles of learning and is, therefore, an excellent way to improve learning

Studying with a pen or highlighter is a visual as well as a tactile/concrete way to improve your concentration. Placing the pen or highlighter in your hand and using it will force you to concentrate more on what you are reading. After you correctly write and recite the material back to yourself one time, do it five or 10 more times to overlearn it.

Be a Selective Learner

Being selective in your math learning will improve your memory. Prioritize the materials you are studying; decide which facts you need to know and which ones you can ignore. Narrow down information into laws and principles that can be generalized. Learn the laws and principles 100 percent.

Example: If you have been given a list of math principles and laws to learn for a test, put each one on an index card. As you go through them, create two piles: a "I already know this" pile and a "I don't know this" pile. Then, study *only* the "I don't know this" pile. Study the "I don't know this" pile until it is completely memorized and understood.

Become an Organizer

Organizing your math material into idea/fact clusters or groups will help you learn and memorize it. Grouping similar material in a problem log or calculator log are examples of categorizing information. Do not learn isolated facts; always try to connect them to other similar material.

Use Visual Imagery

Using mental pictures or diagrams to help you learn the material is especially helpful for the visual learners and those who are right-hemisphere dominant (who tend to learn best by visual and spatial methods). Mental pictures and actual diagrams involve 100 percent of your brain power. Picture the steps to solve difficult math problems in your mind.

> **Example:** Use The Foil Method (see Figure 16) to visually learn how to multiply binomials. Memorize the face until you can sketch it from memory. If you need to use it during a test, you can then sketch the face onto your scratch paper and refer to it.

Make Associations

Association learning can help you remember better. Find a link between new facts and some well-established old facts and study them together. The recalling of old facts will help you remember the new facts and it strengthens a mental connection between the two. Make up your own associations to remember math properties and laws.

> **Remember:** The more ridiculous the association, the more likely you are to remember it.

> **Examples:** When learning the *commutative property*, remember that the word "commutative" sounds like the word "community." A community is made up of different types of people who could be labeled as an "a" group and a "b" group. However, in a community of "a" people and "b" people, it does not matter if we count the "a" people first or the "b" people first; we still have the same total number of people in the community. Thus, a+b=b+a.
>
> When learning the *distributive law of multiplication over addition*, such as a(b+c), remember that "distributive" sounds like "distributor," which is associated with giving out a product. The distributor "a" is giving its products to "b" and "c."

Use Mnemonic devices

The use of mnemonic devices is another way to help you remember. Mnemonic devices are easily remembered words, phrases or rhymes associated with difficult-to-remember principles or facts.

Example: Many students become confused when using the *Order of Operations*. These students mix up the order of the steps in solving a problem, such as dividing instead of first adding the numbers in the parentheses. A mnemonic device to remember the *Order of Operations* is "Please Excuse My Dear Aunt Sally." The first letter in each of the words represents the math function to be completed from the first to the last. Thus, the *Order of Operations* is **P**arentheses (*Please*), **E**xponents (*Excuse*), **M**ultiplication (*My*), **D**ivision (*Dear*), **A**ddition (*Aunt*), and **S**ubtraction (*Sally*). Remember to multiply and/or divide whatever comes first, from left to right. Also, add or subtract whatever comes first, from left to right.

Use Acronyms

Acronyms are another memory device to help you learn math. Acronyms are word forms created from the first letters of a series of words.

Example: FOIL is one of the most common math acronyms. FOIL is used to remember the procedure to multiply two binomials. Each letter in the word FOIL represents a math operation. **FOIL** stands for **F**irst, **O**utside, **I**nside and **L**ast, as it applies to multiplying two binomials such as $(2x+3)(x+7)$. The **F**irst quantity is $2x$ (in the first expression) and x (in the second expression). The **O**utside quantity is $2x$ (in the first expression) and 7 (in the second expression). The **I**nside quantity is 3 (in the first expression) and x (in the second expression). The **L**ast quantity is 3 (in the first expression) and 7 (in the second expression). This results in **F** $((2x)(x)) +$ **O** $((2x)(7)) +$ **I** $((3)(x)) +$ **L** $((3)(7))$.

Do the multiplication to get $2x^2 + 14x + 3x + 21$, which adds up to $2x^2 + 17x + 21$. See Figure 16 (The FOIL Method), p. 192.

How to develop practice tests

Developing a practice test is one of the best ways to evaluate your memory and math skills before being given a real test. You want to find out what you do not know *before* the real test instead of *during* the test. Practice tests should be as real as possible and should include the use time constraints.

You can create a practice test by reworking all the problems, since your last test, that you have recorded in your problem log. Another practice test can be developed using every other problem in the textbook chapter tests. Further, you can use the solutions manual to generate other problems with which to test yourself. You can also use old exams from the previous semester. Check to see if the math lab/ LRC or library has tests on file from previous semesters, or ask your instructor for other tests. For some students, a better way to prepare for a test is the group method.

> **Example:** Hold a group study session several days before the test. Have each student prepare a test with 10 questions. On the back of the test, have listed the answers, worked out step by step. Have each member of the study group exchange his/her test with another member of the group. Once all the tests have been completed, have the author of each test discuss with the group the procedures used to solve those problems.

If group work improves your learning, you may want to hold a group study session at least once a week. Make sure the individual or group test is completed at least three days before the real test.

Completing practice math tests will help you increase testing skills. It will also reveal your test problem weaknesses in enough time for you to learn how to solve the problem before the real test. If you have difficulty with any of the problems during class or after taking the practice test, be sure to see your tutor or instructor.

After taking the practice test(s), you should know what parts you do not understand (and need to study) and what is likely to be on the test. Put this valuable information on one sheet of paper. This information needs to be understood and memorized. It may include formulas, rules or steps to solving a problem.

Use the learning strategies discussed in this chapter to remember this information. A good example of how this information should look is what students might call a mental "cheat sheet." Obviously, you cannot use the written form of this sheet during the real test.

If you cannot take a practice test, put down on your mental cheat sheet the valuable information you will need for the test. Work to understand and memorize your mental cheat sheet. Chapter 10, "How to Improve Your Math Test-Taking Skills," will discuss how to use the information on the mental cheat sheet – *without cheating.*

How to use number sense

When taking your practice tests, you should use "number sense" or estimations to make sure your answer is reasonable. Number sense is like common sense but it applies to math. Number sense is the ability to see if your answer makes sense without using algorithms. (Algorithms are the sequential math steps used to solve problems.) These following two examples demonstrate solving two math problems (from a national math test given to high school students) using algorithms and number sense.

Example One: Solve 3.04 x 5.3. Students use algorithms to solve this problem by multiplying the number 3.04 by 5.3, in sequence. Seventy-two percent of the students answered the problem correctly using algorithms.

Example Two: Estimate the product of 3.04 x 5.3, and answer given the choices below

A)   1.6   C) 160
B)   16   D) 1600

Only 15 percent of the students chose "B," which is the correct answer. Twenty-eight percent of the students chose "A." Using *estimating* to solve the answer, a whopping 85 percent of the students got the problem wrong.

These students were incorrectly using their "mental black board" instead of using number sense. In using number sense to answer, you would multiply the numbers to the left of the decimal in each number to get an estimate of the answer. To estimate the answer, then, you would multiply 3 (the number to the left of the decimal in 3.04) by 5 (the number to the left of the decimal in 5.3) and expect the answer to be a little larger than 15.

It appears that the students' procedural processing (the use of algorithms) was good, but when asked to solve a non-routine problem using estimating (which is easier than using algorithms), the results were disappointing.

Another example of using number sense or estimating is in "rounding off."

Example: Solve 48 + 48 by rounding off. Rounding off means mentally changing the number (up or down) to make it more manageable to you, without using algorithms. By rounding off, 48 becomes 50 (easier to work with). 50 + 50 = 100. If the choices for answers were 104, 100, 98 and 96, you would then subtract four from the 100 (since each number was rounded up by two) and you would get 96.

Taking the time to estimate the answer to a math problem is a good way to check your answer.

Another way to use number sense is to check your answer to see if it is reasonable. Many students forget this important step and get the answer wrong. This is especially true of word or story problems.

> **Examples:** When solving a rate-and-distance problem, use your common sense to realize that one car cannot go 500 miles per hour to catch the other car. However, the car could go 50 miles per hour.
>
> The same common-sense rule applies to age-word problems where the age of a person cannot be 150 years, but it could be 15 years.
>
> Further, in solving equations, "x" is *usually* a number that is less than 20. When you solve a problem for "x" and get 50, then this isn't reasonable, and you should recheck your calculations.

Also remember that, when dealing with an equation, you make sure to put the answer back into the equation to see if one side of the equation equals the other. If the two sides are not equal, you have the wrong answer. If you have extra time left over after you have completed a test, you should check answers using this method.

> **Example:** In solving the equation x+3=9, you calculated that x=5. To check your answer, substitute x with 5 and see if the problem works out correctly. 5+3 does not equal 9, so you know you have made a mistake and need to recalculate the problem. The correct answer, by the way, is x=6.

Remember: Number sense is a way to get more math problems correct by estimating your answer to determine if it is reasonable.

Summary

— Remembering what you learn begins with understanding the relationship between receiving (sensing), storing (processing) and retrieving (recalling) information.

— Having enough working memory to recall information from long-term memory to solve a problem is important.

— Transforming working memory into long-term memory is the major memory problem for most students.

— While studying, many students do not complete this memory-shifting process.

— Understanding the stages of memory and using memory techniques can help you store information in long-term memory.

— Common memory techniques include maintaining a good math/ study attitude, using your best learning sense, becoming a selective learner, becoming an organizer, using visual imagery, making associations, using mnemonic devices and using acronyms.

— Developing practice tests can also help you learn where your memory is failing you, and creating practice tests helps you increase your test-taking skills.

— Memory output skills (or recalling long-term memory into working memory) can be improved.

— Use your imagination to adapt these learning techniques to the math material you need to understand and learn.

— One way to improve these skills is to become more automatic with your mental processing of numbers and using number sense.

— These two techniques, plus the use of a calculator, can free up more working memory to help you solve the problem.

Remember: Locating where your memory breaks down and compensating for those weaknesses will improve your math learning.

Assignment for Chapter 8

1. Read Reference C – "Student Learning-Style Information," p. 295.

2. Read Reference D – "10 Ways To Improve Your Memory," p. 307.

3. Explain the difference between short-term memory and working memory.

4. What are your three best memory techniques?

5. Give an example of how you can use association to learn a math principle.

6. Give an example of a mnemonic device that could be used to improve your learning of math principles.

7. How are you going to develop your practice test?

8. Give two applications of number sense examples.

Chapter 9

How to Reduce Math Test Anxiety

Math test anxiety is a common problem for many college students, and it is especially difficult for students who are in developmental courses. Unfortunately, it is also common for students to experience test anxiety *only* in math and not in their other courses.

Mild test anxiety can be a motivational factor, but high test anxiety can cause major problems both in learning and in taking tests. Reducing test anxiety is the key for many students to become successful in math. Such students need to learn the causes of test anxiety and how to reduce the test anxiety that affects their learning and grades.

Several techniques have proved helpful in reducing both math anxiety and math test anxiety. However, reducing math and test anxiety does not guarantee good math grades. It must be coupled with effective study skills and a desire to do well in math.

In this chapter, you will learn:

– how to recognize test anxiety,

— the causes of test anxiety,

— the different types of test anxiety, and

— how to reduce test anxiety.

How to recognize
test anxiety

Test anxiety has existed for as long as tests have been used to evaluate student performance. Because it is so common and because it has survived the test of time, test anxiety has been carefully studied over the last 50 years. Pioneering studies indicate that test anxiety generally leads to low test scores.

At the University of South Florida (Tampa), Dr. Charles Spielberger investigated the relationship between test anxiety and intellectual ability. The study results suggested that anxiety coupled with high ability can improve academic performance; but, anxiety coupled with low or average ability can interfere with academic performance. That is:

Anxiety + High Ability = Improvement
Anxiety + Low or Average Ability = No Improvement

Example: Students of average ability and with low test anxiety had better performance and higher grades than did students of average ability with high test anxiety. However, there are students who make good grades, are in calculus and who still have test anxiety.

Test anxiety is a *learned* response; a person is not born with it. An environmental situation brings about test anxiety. The good news is that, because it is a learned response, it can be *unlearned*. Test anxiety is a special kind of general stress. General stress is considered "strained exertion," which can lead to physical and psychological problems. For a better understanding of general stress, read Reference E — "Stress," p. 311.

Defining Test Anxiety

There are several definitions of test anxiety. One definition states, "Test anxiety is a conditioned emotional habit to either a single terrifying experience, recurring experience of high anxiety, or a continuous condition of anxiety." (Wolpe, 1958)

Another definition of test anxiety relates to the educational system. The educational system develops evaluations which measure one's mental performance, which creates test anxiety. This definition suggests that test anxiety is the *anticipation* of some realistic or nonrealistic situational threat. (Cattell, 1966) The "test" can be a research paper, an oral report, work at the blackboard, a multiple-choice exam, a written essay or a math test.

Math test anxiety is a new concept in education. *Ms. Magazine* (1976) published "Math Anxiety: Why is a Smart Girl Like You Counting on Your Fingers?" and coined the phrase "math anxiety." During the 1970s, other educators began using the terms "mathophobia" and "mathemaphobia" as a possible cause for children's unwillingness to learn math. Additional studies on the graduate level discovered math anxiety was common among adults, as well as children. Educators did not focus on math anxiety as a state of mind, but as a skill deficiency, until the *Ms. Magazine* article appeared.

There are several other definitions of math anxiety:

Math anxiety is the extreme reaction to a very negative attitude toward math. (There is a strong relationship between low math confidence and high math test anxiety; this could be the same concept.) (Fox, 1977)

Math anxiety is the feeling of tension and anxiety that interferes with the manipulation of numbers and the solving of math problems during school tests. (Richardson and Sulnn, 1972)

Math anxiety is a state of panic, helplessness, paralysis and mental disorganization that occurs in some students when required to solve math problems. This discomfort varies in intensity and is the outcome of numerous previous situations. (Tobias, 1986)

One of my students once described math test anxiety as, "Being in a burning house with no way out."

No matter how you define it, math test anxiety is real, and it affects millions of students.

Why Math Tests Create Anxiety

Math anxiety can be divided into two separate anxieties: *Math test* anxiety and *numerical* anxiety. Math test anxiety involves anticipation, completion and feedback of math tests. Numerical anxiety refers to everyday situations requiring working with numbers.

It has been shown that math anxiety exists among many students who usually do not suffer from other tensions. Counselors at a major university reported that one-third of the students who enrolled in behavior therapy programs, offered through counseling centers, had problems with math anxiety. (Sulnn, 1970)

Educators know that math anxiety is a common among college students and is more prevalent in women than in men. They also know that math anxiety frequently occurs in students with a poor high school math background. These students were found to have the greatest amount of anxiety.

Approximately half of the students in college prep math courses (designed for students with inadequate high school math background or low placement scores) could be considered to have math anxiety.

However, math anxiety also occurs in students in high-level math courses, such as college algebra and calculus.

Educators investigating the relationship between anxiety and math have indicated that anxiety contributes to poor grades in math. They also found that simply *reducing* math test anxiety does not guarantee higher math grades.

The causes of test anxiety

The causes of test anxiety can be different for each student, but they can be explained by seven basic concepts (described on the next page).

The most common situation students have reported as a known cause of their math test anxiety is their elementary school math experiences. When asked, many students indicated that they were made fun of when trying to solve math problems at the chalkboard. When they could not solve the problem, the teacher and/or students would call them "stupid."

Teacher and peer embarrassment and humiliation become the conditioning experience that causes some students' test anxiety. Over the years, this test anxiety is reinforced and still exists. In fact, many math anxious students – now 30 and 40 years old – *still* have extreme fear about working math problems on the board. Some students said that they absolutely refuse go to the board.

The Causes of Test Anxiety

1. Test anxiety can be a learned behavior resulting from the expectations of parents, teachers or other significant people in the student's life.

2. Test anxiety can be caused by the association between grades and a student's personal worth.

3. Test anxiety develops from fear of alienating parents, family or friends due to poor grades.

4. Test anxiety can stem from a feeling of lack of control and an inability to change one's life situation.

5. Test anxiety can be caused by student's being embarrassed by the teacher or other students when trying to do math problems.

6. Test anxiety can be caused by timed tests and the fear of not finishing the test even if one can do all the problems.

7. Test anxiety can be caused by being put in math courses above the student's level of competence.

What are the cause(s) of your test anxiety?

The different types of test anxiety

The two basic types of test anxiety are emotional (educators term this "somatic") and worry (educators term this "cognitive"). Students with high test anxiety have *both* emotional and worry anxiety.

Signs of emotional anxiety are upset stomach, nausea, sweaty palms, pain in the neck, stiff shoulders, high blood pressure, rapid shallow breathing, rapid heartbeat or general feelings of nervousness. As anxiety increases, these feelings intensify. Some students even run to the bathroom to throw up or have diarrhea.

Even though these *feelings* are caused by anxiety, the physical response is real. These feelings and physical inconveniences can affect your concentration, your testing speed, and it can cause you to completely "draw a blank."

Worry anxiety causes the student to think about failing the test. These negative thoughts can happen either before or during the test. This negative "self-talk" causes students to focus on their anxiety instead of recalling math concepts.

The effects of test anxiety range from a "mental block" on a test to avoiding home work. One of the most common side effects of test anxiety is getting the test and immediately forgetting information that you know. Some students describe this event as having a "mental block," "going blank" or indicating that the test looks like Greek.

After five or 10 minutes into the test, some of these students can refocus on the test and start working the problems. They have, however, lost valuable time. For other students, anxiety persists throughout the test and they cannot recall the needed math information. It is only after they walk out of the door that they can remember how to work the problems.

Sometimes math anxiety does not cause students to "go blank" but it slows down their mental processing speed. This means it takes longer to recall formulas and concepts and to work problems. The result is frustration and loss of time, leading to *more* anxiety. Since, in most cases, math tests are *speed* tests (those in which you have a certain amount of time to complete the test), you may not have enough time to work all the problems or to check the answers if you have mentally slowed down. The results is a lower test score, because, even though you knew the material, you did not complete all of the questions before test time ran out.

Not using all of the time allotted for the test is another problem caused by test anxiety. Students know that they should use all of the test time to check their answers. In fact, math is one of the few subjects in which you can check test problems to find out if you have the problems correct. However, most students do not use all of the test time, and this results in lower test scores. Why?

Students with high test anxiety do not want to stay in the classroom. This is especially true of students whose test anxiety increases as the test progresses. The test anxiety gets so bad that they would rather leave early and receive a lower grade than to stay in that "burning house."

Students have another reason for leaving the test early. The fear of what the instructor and other students will think about them for being the last one to hand in the test. These students refuse to be the last group to finish the test because they feel the instructor or other students will think they are dumb. This is middle-school thinking, but the feelings are still real — no matter the age of the student. These students do not realize that some students who turn in their tests first fail, while many students who turn in their tests last make "A's" and "B's."

Another effect of test anxiety relates to completing homework assignments. Students who have high test anxiety may have difficulty starting or completing their math homework. Doing home work reminds some students of their learning problems in math. More specifically, it reminds them of their previous math failures, which causes further anxiety. This anxiety can lead to total homework avoidance or "approach-avoidance" behavior.

Total homework avoidance is called procrastination. The very thought of doing their homework causes these students anxiety, which causes them to put off tackling their homework. This makes them feel better for a short amount of time — *until test day.*

> **Example:** Some students begin their homework and work some problems successfully. They then get stuck on an problem that causes them anxiety, so they take a break. During their break the anxiety disappears until they start doing their homework again. Doing their homework causes more anxiety which leads to another break. The breaks become more frequent. Finally, the student ends up taking one long break and not doing the homework. Quitting, to them, means *no more anxiety* until the next homework assignment.

The effects of math test anxiety can be different for each student. Students can have several of the mentioned characteristics that can interfere with math learning and test taking. However, there are certain myths about math that each student needs to know. Review Figure 17 (The 12 Myths About Test Anxiety), p. 212, to see which ones you believe. If you have test anxiety, which of the mentioned characteristics are true of you?

How to reduce test anxiety

To reduce math test anxiety, you need to understand both the relaxation response and how negative self-talk undermines your abilities.

Relaxation Techniques

The relaxation response is any technique or procedure that helps you to become relaxed and will take the place of an anxiety response. Someone simply telling you to relax or even telling yourself to relax, however, without proper training, does little to reduce your test anxiety. There are both short-term and long-term relaxation response tech-

Figure 17

The 12 Myths About Test Anxiety

1. Students are born with test anxiety.

2. Test anxiety is a mental illness.

3. Test anxiety cannot be reduced.

4. Any level of test anxiety is bad.

5. All students who are not prepared have test anxiety.

6. Students with test anxiety cannot learn math.

7. Students who are well prepared will not have test anxiety.

8. Very intelligent students and students taking high-level courses, such as calculus, do not have test anxiety.

9. Attending class and doing my homework should reduce all my test anxiety.

10. Being told to relax during a test will make you relaxed.

11. Doing nothing about test anxiety will make it go away.

12. Reducing test anxiety will guarantee better grades.

niques which help control emotional (somatic) math test anxiety. These techniques will also help reduce worry (cognitive) anxiety. Effective *short-term* techniques include The Tensing and Differential Relaxation Method and The Palming Method.

Short-Term Relaxation Techniques

The Tensing and Differential Relaxation Method

The Tensing and Differential Relaxation Method helps you relax by tensing and relaxing your muscles all at once. Follow these procedures while you are sitting at your desk before taking a test:

1. Put your feet flat on the floor.

2. With your hands, grab underneath the chair.

3. Push down with your feet and pull up on your chair at the same time for about five seconds.

4. Relax for five to 10 seconds.

5. Repeat the procedure two to three times.

6. Relax all your muscles except the ones that are actually used to take the test.

The Palming Method

The palming method is a visualization procedure used to reduce test anxiety. While you are at your desk before or during a test, follow these procedures:

1. Close and cover your eyes using the center of the palms of your hands.

2. Prevent your hands from touching your eyes by resting the lower parts of your palms on your cheekbones and placing your fingers on your forehead. Your eyeballs must not be touched, rubbed or handled in any way.

3. Think of some real or imaginary relaxing scene. Mentally visualize this scene. Picture the scene as if you were actually there, looking through your own eyes.

4. Visualize this relaxing scene for one to two minutes.

Practice visualizing this scene several days before taking a test and the effectiveness of this relaxation procedure will improve.

Side One of the audio cassette, *How to Reduce test Anxiety* (Nolting, 1986), further explains test anxiety and discusses these and other short-term relaxation response techniques. Short-term relaxation techniques can be learned quickly but are not as successful as the long-term relaxation technique. Short-term techniques are intended to be used while learning the long-term technique.

Long-Term Relaxation Techniques

The Cue-Controlled Relaxation Response Technique is the best long-term relaxation technique. It is presented on Side Two of the audio cassette, *How To Reduce Test Anxiety* (Nolting, 1986). Cue-controlled relaxation means you can induce your own relaxation based on repeating certain cue words to yourself. In essence, you are taught to relax and then silently repeat cue words, such as "I am relaxed."

After enough practice, you can relax during math tests. The Cue-Oriented Relaxation Technique has worked with thousands of students. For a better understanding of test anxiety and how to reduce it, listen to *How to Reduce Test Anxiety* (Nolting, 1986).

Negative Self-Talk

Negative self-talk is a form of worry (cognitive) anxiety. This type of worrying can interfere with your test preparation and can keep you from concentrating on the test. Worrying can motivate you to study, but too much worrying may prevent you from studying at all.

Negative self-talk is defined as the negative statements you tell yourself before and during tests. Negative self-talk causes students to lose confidence and to give up on tests. Further, it can give you an inappropriate excuse for failing math and cause you to give up on learning math.

Students need to change their negative self-talk to positive self-talk without making unrealistic statements.

Positive self-statements can improve your studying and test preparation. During tests, positive self-talk can build confidence and decrease your test anxiety. These positive statements (see examples), as well as others, can help reduce your test anxiety and improve your grades. Some more examples of positive self statements are on the cassette tape *How to Reduce Test Anxiety* (Nolting, 1986). Before the test, make up some positive statements to tell yourself.

Examples of Negative Self-talk:

"No matter what I do, I will not pass this course."

"I failed this course last semester, and I will fail it again."

"I am no good at math, so why should I try?"

"I cannot do it; I cannot do the problems, and I am going to fail this test."

"I have forgotten how to do the problems, and I am going to fail."

"I am going to fail this test and never graduate."

"If I can't pass this test, I am too dumb to learn math and will flunk out."

Examples of Positive Self-talk:

"I failed the course last semester, but I can now use my math study skills to pass this course."

"I went blank on the last test, but I now know how to reduce my test anxiety."

"I know that my poor math skills are due to poor study skills, not my own ability, and since I am working on my study skills, my math skills will improve."

"I know that, with hard work, I will pass math."

"I prepared for this test and will do the best I can. I will reduce my test anxiety and use the best test-taking procedures. I expect some problems will be difficult, but I will not get discouraged."

"I am solving problems and feel good about myself. I am not going to worry about that difficult problem; I am going to work the problems that I can do. I am going to use all the test time and check for careless errors. Even if I do not get the grade I want on this test, it is not the end of the world."

Summary

– General test anxiety is a learned behavior developed by having emotional and/or worry (somatic and/or cognitive) responses during previous tests.

– General test anxiety is a fear of any type of test.

– Math test anxiety is a subclass of general test anxiety that is specific to one subject area.

– Math test anxiety, like general test anxiety, can decrease your ability to perform on tests.

– Your ability is decreased by blocked memory and an urgency to leave the test room before checking all your answers.

– To reduce test anxiety, you must practice relaxations techniques and develop your own positive self-talk statements.

– Reducing your math test anxiety does not guarantee success on tests; first you have to know the appropriate material to recall during the test and have good test-taking skills.

– To substantially reduce high test anxiety, follow the instructions presented on Side Two of *How to Reduce Test Anxiety* audio cassette tape (Nolting, 1986).

Remember: Reducing test anxiety will not happen overnight – it will take time and practice.

Assignment for Chapter 9

1. Read Reference E — "Stress," p. 311.

2. Listen to the tape *How to Reduce Test Anxiety* (Nolting, 1986).

3. From Side One of the *How To Reduce Test Anxiety* audio cassette tape, describe the short-term relaxation technique that works best for you. Practice your short-term relaxation technique.

4. List two positive self-talk statements that you will use during homework.

5. List two positive self-talk statements you will use on the next test.

6. Practice The Cue-controlled Relaxation Technique every day until it takes you two minutes or less to relax before taking a test.

Chapter 10

How to Improve Your Math Test-Taking Skills

Taking a math test is different from taking tests in other subjects. Math tests not only require you to recall the information, you must apply the information. Multiple-choice tests, for example, usually test you on recall, and if you do not know the answer, you can guess.

Math tests build on each other where history tests often do not test you on previous material. Most math tests are speed tests where the faster your are, the better grade you can receive, while most social science tests are designed for everyone to finish.

Math test preparation and test-taking skills are different from preparation and skills needed for other tests. You need to have a test-taking plan and a test-analysis plan to demonstrate your total knowledge on math tests. Students with these plans make better grades compared to students without them. Math instructors want to measure your math knowledge, not your poor test-taking skills.

In this chapter your will learn:

– why attending class and doing your homework may not be enough to pass,

— the general pretest rules,

— the 10 steps to better test-taking,

— the six types of test-taking errors, and

— how to prepare for the final exam.

Why attending class and doing your homework may not be enough to pass

Most students and some instructors believe that attending class and doing all their homework ensures an "A" or "B" on tests. This is far from true. Doing all the homework and getting the correct answers is very different in many ways from taking tests:

1. While doing homework, there is little anxiety. A test situation is just the opposite.

2. You are not under a time constraint while doing your homework; you may have to complete a test in 55 minutes or less.

3. If you get stuck on a homework problem, your textbook and notes are there to assist you. This is not true for most tests.

4. Once you learn how to do several problems in a homework assignment, the rest are similar. In a test the problems are all in random order.

5. In doing homework, you have the answers to at least half the problems in the back of the text and answers to all the problems

in the solutions guide. This is not true for tests.

6. While doing homework, you have time to figure out how to correctly use your calculator. During the test you can waste valuable time figuring out how to use your calculator.

7. When doing homework, you can call your study buddy or ask the tutor for help, something which you cannot do on the test.

Do not develop a false sense of security by believing you can make an "A" or "B" by just doing your homework. Tests measure more than just your math knowledge.

The general pretest rules

General rules are important when taking any type of test:

1. *Get a good night's sleep before taking a test.* This is true for the ACT, the SAT and your math tests. If you are going to cram all night and imagine you will perform well on your test with three to four hours of sleep, you are wrong. It would be better to get seven or eight hours sleep and be fresh enough to use your memory to recall information needed to answer the questions.

2. *Start studying for the test at least three days ahead of time.* Make sure you take a practice test to find out, before the test, what you do not know. Review your problem log and work the problems. Review the concept errors you made on the last test. (How to identify and correct your concept errors will be discussed later on in this chapter.) Meet with the instructor or tutor for

help on those questions you cannot solve.

3. *Review only already learned material the night before a test.*

4. *Make sure you know all the information on your mental cheat sheet, but do not use it on the test.* Review your notebook and glossary to make sure you understand the concepts. Work a few problems and recall the information on your mental cheat sheet right before you go to bed. Go directly to bed; do not watch television, listen to the radio or party. While you are asleep, your mind will work on and remember the last thing you did before going to sleep.

5. *Get up in the morning at your usual time and review your notes and problem log.* Do not do any new problems but make sure your calculator is working.

The 10 steps to better test-taking

Once you begin a test, follow the 10 steps to better test-taking, below:

Step 1 – *Use a memory data dump.* Upon receiving your test, turn it over and write down the information that you put on your mental cheat sheet. Your mental cheat sheet has now turned into a mental list and writing down this information is not cheating. Do not put your name on it, do not skim it, just turn it over and write down those facts, figures and formulas from your mental cheat sheet or other information you might not remember during the test. This is called your *first memory data dump.* The data dump provides memory cues for test questions.

> **Example:** It might take you a while to remember how to do a coin-word problem. However, if you had immediately turned your test over and written down different ways of solving coin-word problems, it would be easier to solve the coin-word problem.

Step 2 – *Preview the test.* Put your name on the test and start previewing. Previewing the test requires you to look through the entire test to find different types of problems and their point values. Put a mark by the questions that you can do without thinking. These are the questions that you will solve first.

Step 3 – *Do a second memory data dump.* The secondary data dump is for writing down material that was jarred from your memory while previewing the test. Write this information on the back of the test.

Step 4 – *Develop a test progress schedule.* When you begin setting up a test schedule, determine the point value for each question. You might have some test questions that are worth more points than others.

In some tests, word problems are worth five points and other questions might be worth two or three points. You must decide the best way to get the most points in the least amount of time. This might mean working the questions worth two to three points first and leaving the more difficult word problems for last.

Decide how many problems should be completed halfway though the test. You should have more than half the problems completed by that time.

Step 5 – *Answer the easiest problems first.* Solve, in order, the problems you marked while previewing the test. Then, review the answers to see if they make sense. Start working through the test as fast as you can while being accurate. Answers should be reasonable.

> **Example:** The answer to a problem of trying to find the area of a rectangle cannot be negative, and the answer to a land-rate-distance problem cannot be 1,000 miles per hour.

Clearly write down each step to get partial credit, even if you end up missing the problem. In most math tests, the easier problems are near the beginning of the first page; you need to answer them efficiently and quickly. This will give you both more time for the harder problems and time to review.

Step 6 – *Skip difficult problems.* If you find a problem that you do not know how to work, read it twice and automatically skip it. Reading it twice will help you understand the problem and put it into your working memory. While you are solving other problems, your mind is still working on that problem. Difficult problems could be the type of problem you have never seen before or a problem in which you get stuck on the second or third step. In either case, skip the problem and go on to the next one.

Step 7 – *Review the skipped problems.* When working the skipped problems, think how you have solved other, similar problems as a cue to solving the skipped ones. Also try to remember how the instructor solved that type of problem on the board.

While reviewing skipped problems, or at any other time, you may have the "Ah, ha!" response. The "Ah, ha!" response is your remembering how to do a skipped problem. Do not wait to finish your current problem. Go to the problem on which you had the "Ah ha" and finish that problem. If you wait to finish your current problem, your "Ah, ha!" response could turn into an "Oh, no!" response.

Step 8 – *Guess at the remaining problems.* Do as much work as you can on each problem, even if it is just writing down the first step. If you cannot write down the first step, rewrite the problem. Sometimes rewriting the problem can jar your memory enough to do the first step or the entire problem. If you leave the problem blank, you will get a zero. Do not waste too much time on guessing or trying to work the problems you cannot do.

Step 9 – *Review the test.* Look for careless errors or other errors you may have made. Students usually lose two to five test points on errors that could have been caught in review. Do not talk yourself out of an answer just because it may not look right. This often happens when an answer does not come out even. It is possible for the answer to be a fraction or decimal.

Remember: Answers in math do not have "dress codes." Research reveals that the odds of changing a right answer to a wrong answer are greater than the odds of changing a wrong answer to a right one.

Step 10 – *Use all the allowed test time.* Review each problem by substituting the answer back into the equation or doing the opposite function required to answer the question. If you cannot check the problem by the two ways mentioned, rework the problem on a separate sheet of paper and compare the answers. Do not leave the test room unless you have reviewed each problem two times or until the bell rings.

Remember: There is no prize for handing your test in first, and students who turn their papers in last do make "A's."

Stapling your scratch paper to the math test when handing it in has several advantages:

– If you miscopied the answer from the scratch paper, you will probably get credit for the answers.

—   If you get the answer incorrect due to a careless error, your work
    on the scratch paper could give you a few points.

—   If you do get the problem wrong, it will be easier to locate the
    errors when the instructor reviews the test. This will prevent you
    from making the same mistakes on the next math test.

    Remember: Handing in your scratch paper may get you ex-
    tra points or improve your next test score.

The six types of test-taking errors

To improve future test scores, you must conduct a test analysis of
previous tests. In analyzing your tests, you should look for the following kinds of errors.

1. misread-direction errors

2. careless errors

3. concept errors

4. application errors

5. test-taking errors

6. study errors

Students who conduct math-test analyses will improve their total test scores.

Misread-direction errors occur when you skip directions or mis-understand directions, but do the problem anyway.

Examples: 1] You have this type of problem to solve: $(x+1)(x+1)$. Some students will try to solve for x, but the problem only calls for multiplication. You would solve for x only if you have an equation such as $(x+1)(x+1)=0$.

2] Another common mistake is not reading the directions before doing several word problems or statistical problems. All too often, when a test is returned, you find only three out of the five problems had to be completed. Even if you did get all five of them correct, it costs you valuable time which could have been used obtaining additional test points.

To avoid misread-direction errors, read all the directions. If you do not understand them, ask the instructor for clarification.

Careless errors are mistakes made which you could catch automatically upon reviewing the test. Both good and poor math students make careless errors. Such errors can cost a student the difference of a letter grade on a test.

Examples: 1] *Dropping the sign*: $-3(2x)=6x$, instead of $-6x$, which is the correct answer. 2] *Not simplifying your answer:* Leaving $(3x - 12)/3$ as your answer instead of simplifying it to $x - 4$. 3] *Adding fractions*: $1/2 + 1/3 = 2/5$, instead of $5/6$, which is the correct answer. 4] *Word problems*: X = 15 instead of the "student had 15 tickets."

However, many students want all their errors to be careless errors. This means that the students *did* know the math, but simply made silly mistakes. In such cases, I ask the student to solve the problem immediately, while I watch.

If the student can solve the problem or point out his/her mistake in a few seconds, it is a careless error. If the student cannot solve the problem immediately, it is not a careless error and is probably a concept error.

When working with students who make careless errors, I ask them two questions: First, "How many points did you lose due to careless errors?" Then I follow with, "How much time was left in the class period when you handed in your test?" Students who lose test points to careless errors are giving away points if they hand in their test papers before the test period ends.

To reduce careless errors, you must realize the types of careless errors made and recognize them when reviewing your test. If you cannot solve the missed problem immediately, it is not a careless error. If your major error is not simplifying the answer, review each answer as if it were a new problem and try to reduce it.

Concept errors are mistakes made when you do not understand the properties or principles required to work the problem. Concept errors, if not corrected, will follow you from test to test, causing a loss of test points.

Examples: Some common concept errors are not knowing:

$(-)(-)x = x$, *not "-x"*

$-1(2)>x(-1) = 2<x$, *not "2>x"*

$5/0$ is undefined, *not "0"*

$(a+x)/x$ *is not reduced to "a"*

The order of operations

Concept errors must be corrected to improve your next math test score. Students who have numerous concept test errors will fail the next test, and the course, if concepts are not understood. Just going back to rework the concept error problems is not good enough. You must go back to your textbook or notes and learn why you missed those types of problems, not just the one problem itself.

The best way to learn how to work those types of problems is to set up a concept-problem error page in the back of your notebook. Label the first page "Test One Concept errors." Write down all your concept errors and how to solve the problems. Then, work five more problems which use the same concept. Now, in your own words, write the reasons that you *can* solve these problems.

If you cannot write the concept in your own words, you do not understand it. Get assistance from your instructor if you need help finding similar problems using the same concept or cannot understand the concept. Do this for every test.

Application errors occur when you know the concept but cannot apply it to the problem. Application errors usually are found in word problems, deducing formulas (such as the quadratic equation) and graphing. Even some better students become frustrated with application errors; they understand the material but cannot apply it to the problem.

To reduce application errors, you must predict the type of application problems that will be on the test. You must then think through and practice solving those types of problems using the concepts.

> **Example:** If you must derive the quadratic formula, you should practice doing that backward and forward while telling yourself the concept used to move from one step to the next.

Application errors are common with word problems. When solving word problems, look for the key phrases displayed in Figure 13 (Translating English Words into Algebraic Expressions), p. 151, to help you set up the problem. After completing the word problem, reread the question to make sure you have applied the answer to the intended question. Application errors can be avoided with appropriate practice and insight.

Test-taking errors apply to the specific way you take tests. Some students consistently make the same types of test-taking errors. Through recognition, these bad test-taking habits can be replaced by good test-taking habits. The result will be higher test scores. The list that follows includes the test-taking errors which can cause you to lose many points on an exam.

1. *Missing more questions in the first third, second third or last third part of a test* is also considered a test-taking error.

   Missing more questions in the first third of a test could be caused by carelessness when doing easy problems or from test anxiety.

   Missing questions in the last part of the test could be due to the fact that the last problems are more difficult than the earlier questions, or due to increasing your test speed to finish the test.

   If you consistently miss more questions in a certain part of the test, use your remaining test time to review that section of the test first. This means you may review the last part of your test first.

2. *Not completing a problem to its last step* is another test-taking error. If you have this bad habit, review the last step of the test problem first, before doing an in-depth test review.

3. *Changing test answers from correct ones to incorrect ones* is a problem for some students. Find out if you are a good or bad answer-changer by comparing the number of answers changed to correct and to incorrect answers. If you are a bad answer-changer, write on your test, "Don't change answers." Change answers only if you can prove to yourself or the instructor that the changed answer is correct.

4. *Getting stuck on one problem and spending too much time on it* is another test-taking error. You need to set a time limit on each problem before moving to the next problem. Working too long on a problem without success will increase your test anxiety and waste valuable time that could be used in solving other problems or in reviewing your test.

5. *Rushing through the easiest part of the test and making careless*

errors is a common test-taking error for the better student. If you have the bad habit of getting more points taken off for the easy problems than for the hard problems, first review the easy problems first and later review the hard problems.

6. *Miscopying an answer from your scratch work to the test* is an uncommon test-taking error, but it does cost some students points. To avoid these kinds of errors, systematically compare your last problem step on scratch paper with the answer written on the test. In addition, always hand in your scratch work with your test.

7. *Leaving answers blank* will get you zero points. If you look at a problem and cannot figure out how to solve it, do not leave it blank. Write down some information about the problem, rewrite the problem or try to do at least the first step.

    Remember: Writing down the first step of a problem is the key to solving the problem and obtaining partial credit.

8. *Answering only the first step of a two-step problem* causes problems for some students. These students get so excited when answering the first step of the problem that they forget about the second step. This is especially true on two-step word problems. To correct this test-taking error, write "two" in the margin of the problem. That will remind you that their are two steps or two answers to this problem.

9. *Not understanding all the functions of your calculator* can cause major testing problems. Some students only barely learn how to use the critical calculator functions. They then forget or have to relearn how to use their calculator, which costs test points and test time. Do not wait to learn how to use your calculator on the test. Overlearn the use of your calculator *before* the test.

10. *Leaving the test early without checking all your answers.* Do

not worry about the first person who finishes the test and leaves. Many students start to get nervous when students start to leave after finishing the test. This can lead to test anxiety, mental blocks and loss of recall.

According to research, the first students finishing the test do not always get the best grades. It sometimes is the exact opposite. Ignore the exiting students, and always use the full time allowed.

Make sure you follow the 10 steps to better test-taking, pp. 222-25. Review your test-taking procedures for discrepancies in following the 10 steps to better test-taking. Deviating from these proven 10 steps will cost you points.

Study error, the last type of mistake to look for in test analysis, occurs when you study the wrong type of material or do not spend enough time on pertinent material. Review your test to find out if you missed problems because you did not practice that type of problem or because you did practice it but forgot how to do it during the test. Study errors will take some time to track down. But correcting study errors will help you on future tests.

Most students, after analyzing one or several tests, will recognize at least one major, common test-taking error. Understanding the effects of this test-taking error should change your study techniques or test-taking strategy.

Example: If there are seven minutes left in the test, should you review for careless errors or try to answer those two problems you could not totally solve? This is a trick question. The real question is, "Do you miss more points due to careless errors or concept errors, or are the missed points about even?" The answer to this question should determine how you will spend the last minutes of the test. If you missed more points due to careless errors or missed about the same number of points due to careless/concept errors, review for careless errors.

Careless errors are easier to correct than are concept errors. However, if you made very few or no careless errors, you should be working on those last two problems to get the greatest number of test points. Knowing your test-taking errors can add more points to your test by changing your test-taking procedure.

Sometimes math tests have objective test questions. Objective test questions are in the form of true/false, multiple choice and matching. To better understand the procedures in taking an objective test, read Reference H – "Taking an Objective test," p. 325.

Very few math tests have essay questions. However, improving the skills required to take an essay test could give you more time for studying math. Read Reference G – "Answering an Essay Test," p. 321.

How to prepare
for the final exam

The first day of class is when you start preparing for the final exam. Look at the syllabus or ask the instructor if the final exam is cumulative. A cumulative exam covers everything from the first chapter to the last chapter. Most math final exams are cumulative.

The second question you should ask is if the final exam is a departmental exam or if it is made up by your instructor. In most cases, departmental exams are more difficult and need a little different preparation. If you have a departmental final, you need to ask for last year's test and ask other students what their instructors say will be on the test.

The third question is, how much will the final exam count? Does it carry the same weight as a regular test or, as in some cases, will it count a third of your grade? If the latter is true, the final exam will usually make a letter grade difference on your final grade. The final exam could also determine if you pass or fail the course. Knowing this information before the final exam will help you prepare.

Preparing for the final exam is similar to preparing for each chapter test. You must create a pretest to discover what you have forgotten. You can use questions from the textbook chapter tests or questions from your study group.

Review the concept errors that you recorded in the back of your
notebook labeled "Test One," "Test Two," etc. Review your problem
log for questions you consistently miss. Even review material that you
knew for the first and second test but which was not used on any
other tests. Students forget how to work some of these problems.

Make sure to use the 10 steps to better test-taking, pp. 222-25,
and the information gained from your test analysis. Use all the time
on the final exam, because you could improve your final grade by a
full letter if you make an "A" on the final.

Summary

- Just completing your homework and attending class does not guarantee that you will pass math.

- Do not cram for your math tests.

- Improving your math test-taking skills begins with completing a practice math test several days before the actual exam. This pre-test can help you locate math areas that need improvement.

- Follow the general principles of good test-taking before each math test.

- Weakness can be corrected by reviewing homework or obtaining help from your instructor or tutor.

- Follow the 10 steps to better test-taking, pp. 222-25, to obtain the greatest number of test points in the least amount of time.

- Make sure you develop a mental cheat sheet before each test for your memory data dump.

- After your fist major test you need to complete a test analysis to learn from your mistakes and to increase test points on the next exam.

- Decide before each test if you are going to spend the last few minutes of each test checking for careless errors or finishing problems you left incomplete.

- Without conducting a test analyses, you will probably continue to make the same old test errors and lose valuable test points.

- You need to start preparing for the final exam the first day of class.

- Ask if the final exam is going to be made up by the instructor or if it a departmental exam.

- Find out how much the final exam counts for the final grade and if it is cumulative.

- To be prepared for each test and the final exam, you must keep up with your problem log and concept errors from each test.

- Before each test, make sure you review your problem log and the concept errors from your previous tests. This type of review will make it easier when preparing for the final exam.

- When preparing for the final exam, make sure to take a pretest developed by either the instructor or yourself. This pretest must be taken under the same timed conditions as the final exam.

- You must have enough time after taking the pretest to learn how to solve the missed problems before the final exam.

- To learn more about math test-taking, read Reference F – "Studying for Exams," p. 317, and listen to the cassette tape, *How to Ace Tests* (available from your college bookstore or by mail order by using the order form in the back of this book).

Remember: It may not be fair, but students with good math test-taking skills can score higher on tests compared to students who have more math knowledge.

Assignment for Chapter 10

1. Read Reference H – "Taking an Objective Test," p. 325.

2. Read Reference G – "Answering an Essay Test," p. 321.

3. Read Reference F – "Studying For Exams," p. 317.

4. Listen to the cassette tape, *How to Ace Tests* (Nolting, 1987).

5. How can you correctly complete all the homework assignments and still not score high on math tests?

6. How can you use a study group to help you prepare for a math test?

7. List the 10 steps to better test-taking.

8. What did you learn from your test analyses?

9. How can you improve your math test-taking skills?

10. How can you prepare for the final exam?

Chapter 11

Considerations for Students with Disabilities

Community colleges and universities are experiencing a large increase in the number of students with disabilities. Improved special education programs have helped students with disabilities graduate from high school.

Most students with disabilities had special education classes with direct teacher instruction or special teachers helping them in the classroom prior to college. Now, without this special help, many college students with disabilities are having difficulty learning in the classroom.

More than ever before, mature students make up a higher percentage of college students. Many of these mature students had previous learning problems in school but were diagnosed as disabled *after* leaving high school. Other mature students have not yet been diagnosed as disabled.

Students with disabilities need additional math learning skills and accommodations to reach their education potential. This is because, in many situations, students with disabilities will have more difficulty learning math. One reason is that most students with disabilities have poor math study skills just like their non-disabled peers.

Students with disabilities also have processing deficits that impair their learning or testing skills. Good math study skills and appropriate accommodations, such as assigned note-takers and extended testing times, can compensate for some processing deficits.

In this chapter you will learn:

— how to define a learning disability, traumatic brain injury and attention deficit disorder,

— the reasons disabilities cause math-learning problems,

— some special study skills for students with disabilities,

— the appropriate accommodations for students with disabilities, and

— how to make an "Individual College Learning/Testing Plan."

How to define a learning disability, traumatic brain injury and attention deficit disorder

In most cases, students with learning disability (LD), traumatic brain injury (TBI), and attention deficit disorder (ADD) will have difficulty in college courses, especially in math. Students with LD may have difficulty in language processing that affects math learning. In almost every case, students with TBI will have problems in math due to short-term memory problems. Students with ADD may have difficulty with math due to concentration problems.

For these students to improve their math learning, they will need math study skills which will make the most of their cognitive processing strengths and help make up for their processing weaknesses. Even though the suggestions in this chapter are for students with disabilities, students who are not disabled may benefit from the study skills suggestions.

Definition of Learning Disabilities

The term "learning disability" (LD) is used to describe a broad range of neurological dysfunctions. Students with LD have average-to above-average intelligence. A learning disability is often misunderstood because it is invisible. The National Joint Committee of Learning Disabilities (1988) defines LD as:

> Learning disabilities is a general term that refers to a heterogeneous group of disorders manifested by significant difficulties in the acquisition and use of listening, speaking, reading, writing, reasoning, or mathematical abilities. These disorders are intrinsic to the individual, presumed to be due to central nervous system dysfunction, and may occur concomitantly with other handicapping conditions (for example, sensory impairment, serious emotional disturbance) or with extrinsic influences (such as cultural differences, insufficient or inappropriate instructions), they are not the results of those conditions or influences.

New research has indicated that learning disabilities are neurologically based and cannot be cured.

Students with LD may present one or more of these symptoms:

— Difficulty doing the actual calculations

— Reversals of numbers, variables or symbols

— Copying problems incorrectly from line to line or off the board

— Difficulty learning a series of math steps to solve a problem

— Inability to apply math concepts to word problems

— Inability to understand or retain abstract concepts

— Poor organizational skills

— Easily distracted

These problems are consistent throughout the students' lives and do not just happen every once in a while. There are many famous people in the past, however, who would qualify as learning disabled: Leonardo Da Vinci, Woodrow Wilson, General George Patton, Sir Winston Churchill, Albert Einstein, Thomas Edison and Hans Christian Andersen.

More current famous people with LD are: Bruce Jenner, Greg Louganis, Nelson Rockefeller, Cher, Danny Gulber and Tom Cruise. From reading these two groups of talented, exception people, you should now know that learning disabilities are not associated with retardation, slowness, laziness or lack of motivation.

Definition of Traumatic Brain Injuries

There is an increase in the number of students attending college with mild to severe traumatic brain injuries (TBIs). These students have graduated from high school with assistance or are attending college after an accident. In other cases, these students had accidents while in college and are now returning.

A traumatic brain injury is any traumatically induced event that causes a loss of consciousness, memory lost, confusion or disorientation, and which results in a neurological deficit. Some TBI causes are accidents involving automobiles, motorcycles, bicycles and in-line skates. TBIs can also occur from sports such as football and boxing. If the effects of a mild TBI last only one day, the person may not seek medical help or may not be informed about potential future problems.

Recovery is different for each student, and the student may not have a full recovery.

The effects of TBIs can be physical, sensory, cognitive, communicative, academic and social. The physical and sensory problems could result in decreased speed in performing tasks, spatial perception and eye-hand coordination.

The most frequent cognitive problem with students is loss of short-term memory and long-term memory. Communication problems could be difficulty in word-finding and reading comprehension. The academic effect of TBIs may be a wide range of grades in different subjects. Low grades in math may be due to problems in reading, speed and following directions.

Some other effects of TBIs could be deficiencies in attention, concentration, reasoning and time management. However, with proper study skills and accommodations, many students with TBIs can pass math and graduate.

Definition of Attention Deficit Disorder

Attention deficit disorder (ADD) is a disability characterized by inattention, impulsivity and, sometimes, hyperactivity. Until the mid-1980s, ADD was believed to be a childhood disorder because as children mature they learned more socially accepted ways to manage their behavior. However, these children – as adults – continue to experience academic problems associated with ADD.

In fact, two to five million adults are considered to be ADD, and it is no longer certain that children "grow out" of ADD. These students are entering colleges and universities, and they must learn behavior strategies and study skills to help them attain their educational goals.

Their main math learning problems are in the areas of computation, word problems, operations and order of operations, mainly due to auditory short-term memory problems.

The reasons disabilities
cause math-learning problems

Students with learning disabilities, who are having difficulty in learning math, may have a math-learning disability or may have problems with different cognitive processes which affect math learning. Cognitive processes include how students receive information and how they understand the information, but they are not *physical* disabilities.

Processing Disorders

Processing disorders block a student's ability to obtain valuable information to learn math and/or to demonstrate math knowledge. If you do not know your type of processing disorder, contact the office for students with disabilities for this important information.

Visual Disorders

Students with learning disabilities who have *visual processing speed disorders* will have difficulty learning math. Visual processing measures how fast you can read numbers and symbols. Such problems may include slow visual speed of working with understood math symbols and numbers — in other words, the speed with which you can copy down recognizable numbers, symbols and problem steps.

Students with a *visual processing speed deficit* will concentrate so much on copying down the notes that they do not understand the lecture. Such students, when taking a math test, cannot read as fast as the other students and may run out of time. The test will measure their reading speed instead of their math knowledge. Their learning breaks down at the sensory register. See Figure 15 (The Stages of Memory), p. 189.

Visual processing deficiency is the inability to recognize and remember, in sequence, complex math symbols and numbers.

> **Example:** The ability to recognize and remember this polynomial: $4x^2 + 2x + 1$. If you have a visual processing deficiency, you may have difficulty telling the difference between 2, as a factor or an exponent, and between "+" (as a plus sign) and "x" (as a variable).

Mistakes can occur in miscopying notes, misreading the text book or misreading questions. This learning breaks down in the sensory register or short-term memory.

Auditory Disorders

Students with *auditory processing difficulties* have problems telling the difference between certain sounds of words. These students, especially in a large lecture hall, will "miss" some words in a lecture or replace some words with other incorrect words. Students will have difficulty understanding the instructor and taking notes. This learning breaks down in the sensory register.

Memory Problems

Short-Term Memory Problems

Students with short-term-memory problems may have difficulty learning math. Short-term memory difficulties cause problems remembering numbers, symbols or words in their the correct order.

This problem becomes apparent when your math instructor explains the steps when working a problem. You then forget the math problem steps or write down the math problem steps in the wrong order. Either mistake will cause difficulty in understanding the math concepts, recording your notes, doing your homework and taking tests.

Short-term memory problems especially affect solving word problems.

Working-Memory Problems

Students with *working-memory/long-term retrieval problems* will eventually have difficulty in math learning. These students may listen to a math lecture and understand each step as it is explained. However, when the instructor goes back to a previous step discussed several minutes before and asks a question, the student cannot explain the reasons for the steps.

When interruptions occur, these students have difficulty remembering series steps long enough to understand the concept, compared to students with short-term memory problems who forget the step as soon as the instructor explains it.

Students with working-memory problems also have difficulty processing several bits of information at a time. Since their working memory is limited, these students have difficulty recalling a math concept from long-term memory into working memory and, at the same time, doing the arithmetic to solve the problem. Students with working memory deficits also have problems putting information into long-term memory and applying concepts to work math problems.

Long-Term Memory Problems

Some students with LD have difficulty with long-term memory. Long-term memory is the stored-up facts and formulas related to math. If you have this problem, you may be inconsistent when learning new facts or concepts. You may be able to learn how to work fractions one day and a week later have difficulty recalling how to work fractions. You may have poor achievement in math calculations but have average or above-average math reasoning ability. These students' major problem is putting information into long-term memory.

Reasoning Disorders

Another type of learning disability is in reasoning or thinking. This problem usually occurs when abstract reasoning is required to apply some type of math concept from its law or principle to its application.

> **Example:** You may have difficulty applying a formula to a new homework problem. You may remember the concept but not know how to generalize it to other problems. In general, you will have difficulty applying abstract math concepts to homework and test problems.

Some special study skills for students with disabilities

Study skills for students with LD may be the same as for other students, but the skills are essential to their learning. Students with LD usually have organizational problems, including time management, and are, in many cases, easily distracted. While some non-disabled students may have an "internal" time-management clock, most students with LD do not.

How to Help Yourself

You must develop a study schedule (p. 98) and complete a weekly study plan (p. 106) The weekly study plan must be completed each week so you know when to study for what subjects/tests. If you have difficulty developing a study schedule or weekly study plan, see the counselor for disabled students for assistance.

Distractions can occur in the classroom, during tutorial sessions, while doing homework and when taking tests. These distractions interfere with your understanding the instructor's lectures and with your studying. You need to sit in The Golden Triangle of Success (p. 113) and as close to the front of the classroom as is comfortable. Being in the center front of the room will decrease your awareness of distractions and provide you the best place to record lectures.

Remember: Sitting in the right place will improve your learning.

How to Get Help from Others

Tutorial sessions can help students with disabilities. The sessions need to be at a routine time and in a quiet environment. Set up your tutorial sessions at the same time each day. Your tutorial sessions may have to be in the late afternoon when the math lab is not crowded.

Make sure your tutor explains to you "how" and "why" a math process works instead of doing the problem for you. Ask your tutor what changes and what does not change if a part of the process is different. Be able to explain the basic fundamentals of the process. *Now you are learning!*

How to Handle Homework

Homework sessions need to be set up on a weekly basis and at the same time each day. Do not skip any homework sessions. Make sure your college study area for doing your homework is in an isolated location.

Study at home during a quiet time of the day or night and put up a "Do not disturb" sign on your door. When doing your homework, put simple numbers into the problems to see what process is at work. Aim to understand the concepts and formulas.

Make up questions for your instructor and tutor. If it helps, write out class notes and text samples. Check each homework problem for which you do not have the answer.

For visual learners, close your eyes and say each of the steps to yourself. For auditory learners, say each problem step out loud until you know them.

Finish your homework by doing problems you can do, even if you have to rework a few problems you already know how to work.

Some study skills adaptations are based on a student's processing deficit or strength. Some students with visual processing/visual discrimination problems have difficulty doing their homework. Their homework looks like "chicken scratch." They mix up their problem steps in solving math equations with the calculations used to solve the equations.

Example: After working three-fourths of a math problem, they cannot find the equations steps. The calculations are so mixed up with equation steps that they cannot find where they left off working the equation. The problem steps and calculations are scattered all over the page. These students bring in their homework problems to the tutors, but the tutors cannot read the problems. To solve this problem, draw a line down the center of the homework page. On the left side of the page write the problem steps, and on the right side of the page do the math calculations.

This technique keeps the problem steps separate from the calculations, solving the problem of mixing up the problem steps and calculations for homework and tests.

Use Color, Felt-tip Pens and Note Cards as Study Aids

Students with visual discrimination misread math notations and signs in their text and while doing homework. Use a 4x6 index card — with no lines on it — as a reading aid. Move the card line by line down the page text or homework page.

When looking at a problem, you will see the exponents first. Highlight them in yellow. Now look at the entire problem. If needed, rewrite the problem using a felt-tip pen.

Do the same thing when working homework problems. A felt-tip pen makes the math numbers and symbols larger, which makes them easier to read. Work each problem step with the felt-tip pen until the problem is finished. If this is still too difficult, turn your note book sideways and use a column for each number or variable. Using enlarged graphing paper to do your homework can also solve this problem.

Another study-skills technique to help discriminate between math numbers and symbols is to use a ball-point pen that has four colors: red, green, blue and black. When doing math homework, use different colors for variables, numbers, exponents, negative signs or positive signs. This does not mean that you have to use a different color for each part of the math problem.

You can use blue for the parts of the math problem you already know how to work and use the other pen colors for the new parts of the math problem. You can follow each math operation by pen color until the problem is completed. Even though it takes you longer to work the problems this way, it decreases visual processing errors and makes it easier to work the problems.

Some students' mental processing speeds get ahead of their writing speeds, which causes them mentally to lose track when solving problems. There are two ways to solve this problem. One way to reduce mental speed is to write each math problem step in a different color pen. This will reduce the mental problem-solving speed, allowing each problem step to be accurately written.

Another solution is to write, in English, the reasons for doing each math step. This process not only slows down a student's mental processing speed but also helps him/her learn the reasons for each problem step.

Some students have difficulty understanding math until they write down the basic processes and concepts. Writing down the reasons for each step during tutorial sessions, homework or studying can help you understand math. The written information can be checked by your tutor and can be used as a future study guide.

For difficult problems, make up multicolored note cards. These note cards will demonstrate a problem representing a difficult concept worked out step by step. Number the reasons for each step on the bottom of the card. If the problem had four steps, there will be four reasons. Highlight the most difficult step so it will stay in your memory. See Figure 18 (Note Card Check System), p. 252.

Cover up the bottom of the card and recall the reasons for each problem step. Review these cards as often as needed.

Remember: Research has shown that writing can help students go beyond the rote learning of math and improve understanding.

Use a Tape Recorder

Students with short-term memory problems must use a tape recorder with a tape counter during tutorial sessions. When working on a difficult problem, record the tutor's explanation. You must also record yourself rephrasing the reasons for the steps while the tutor listens.

Make any corrections that are necessary and write down the tape counter number next to the problem. Listen to the tape before doing similar homework problems and use the tape as a future study reference.

You can use the same procedure when going over your test with the instructor.

Use a "Focusing" Device

Example: Simplify $(-11y + 3y) - (-7y - 5y-4)$. Students look at the total problem, become confused and do not know where to start. To solve this problem, take a piece of cardboard and cut a small rectangle in the center. Remove the small rectangle which leaves a rectangular hole in the cardboard. Lay the piece of cardboard over the first part of the problem. This allows the student to isolate different factors or signs and concentrate on one part of the problem at a time. This eliminates most of the visual confusion caused by a learning disability.

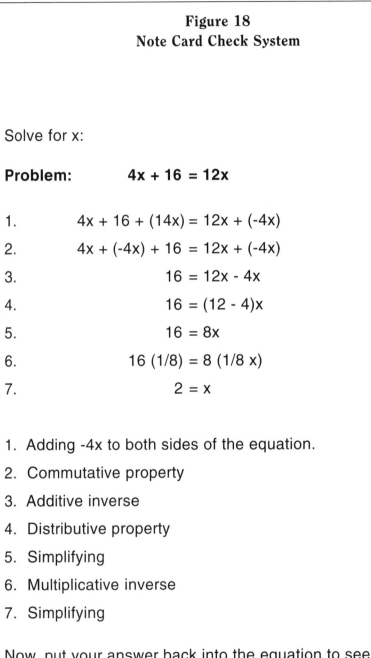

Figure 18
Note Card Check System

Solve for x:

Problem: **4x + 16 = 12x**

1.   4x + 16 + (14x) = 12x + (-4x)
2.   4x + (-4x) + 16 = 12x + (-4x)
3.   16 = 12x - 4x
4.   16 = (12 - 4)x
5.   16 = 8x
6.   16 (1/8) = 8 (1/8 x)
7.   2 = x

1. Adding -4x to both sides of the equation.

2. Commutative property

3. Additive inverse

4. Distributive property

5. Simplifying

6. Multiplicative inverse

7. Simplifying

Now, put your answer back into the equation to see if the right side equals the left side.

Handling Word Problems

Some students with language problems have difficulty understanding and working word problems. Students with LD need to be taught how to break down the word problem into two parts. First, write down the word phrases for each part of the word problem and put the algebraic expressions under each word phrase.

> **Example:** If the word phrase was "a number decreased by seven," write "x − 7" under that word phrase. Use the information in Figure 13 (Translating English Words Into Algebraic Expressions), p. 151, to learn how to write the algebraic expression for word problems.

The "Shift Method" for Graphing Problems

Doing math graphing problems is another challenge. A method for learning graphing is using the "shift method." Rather than merely plotting points and drawing a line though them, develop a concept model about parabolas.

Have your tutor draw a simple parabola on an enlarged graph using the standard parabola equation of $y = (x − h)^2$. Remember that the positive or negative sign inside the parenthesis determines if the parabola is to the left or right of the "y" axis. A negative sign shifts the parabola to the right, and a positive sign shifts the equation to the left.

To better understand the graphing of a parabola, you can draw a basic parabola and then change pen color and draw a new parabola with a shift to the right, change pen color again and draw another parabola with shift to the left.

By looking at the sign of "h," you can tell on which side of the "y" axis to put the graph. Put these graphs on index cards, with different pen colors, to be understood and memorized.

The appropriate accommodations
for students with disabilities

Under Section 504 of the Rehabilitation Act and the Americans with Disabilities Act, students with disabilities are entitled to certain accommodations based on their disabilities. Accommodations for students with disabilities are provided to allow the same access to course material as non-disabled students.

Testing accommodations are provided to make the student with disabilities equivalent to other students. In this way, the disabled students' knowledge, rather than their disability, is tested.

Learning and testing accommodations are based on the students' processing disorders. The mentioned accommodations are commonly suggested for students with learning disabilities, but additional accommodations may also be needed. Some common testing accommodations are extended test times and testing in a private room.

Visual Processing and/or Visual Speed Disorders

Students with visual processing and/or visual speed disorders can visually process materials five to 10 times slower than non-disabled students. These students have major note-taking and test-taking problems. Three lecture accommodations are note-takers, tape recorders and large-print handouts. These students may need only one or all of these learning accommodations. Testing accommodations for these students would include extended test time, enlarged test print and/or color-coded tests.

Auditory Processing and/or Short-Term Memory Disorders

Students with auditory processing disorders, including short-term memory and auditory processing, have difficulty understanding a math lecture. If you have a short-term memory problem, you may not remember the math steps in their correct order or you may forget certain problem steps. You may also get words mixed up with other words. Learning accommodations would be note-takers, tape recorders, close physical proximity to the instructor and video tapes. Students will also need extended test times, because they have to read the problems several times to understand them.

Working Memory and Long-Term Memory/Reasoning Problems

In most cases, students with working memory and long-term memory/reasoning problems will have extreme difficulty learning math. These students usually have poor organizational skills, poor problem-solving skills and difficulty understanding causal relationships. The difficulties which cause the most problems are poor abstract reasoning and difficulty generalizing from one experience/idea to new situations.

These students may have a 3.5 GPA but are failing math. Learning accommodations for LD students with thinking/reasoning problems can be extensive. Most of these students will need a combination of note-takers, tape recorders, handouts, math video tapes, tutors and calculators. Testing accommodations may include calculators, extended test time, enlarged tests, alternate test forms and, in some cases, formula/fact sheets.

Since scientific and graphing calculators are becoming very complicated to use, students may need calculator note cards to help them remember certain key strokes. Calculator note cards are individual 3x5 cards which have examples of they key strokes needed for doing certain calculator functions.

> **Example:** A calculator note card may have listed the key strokes needed to graph two intersecting lines.

The calculator note cards make it possible to test the student's knowledge of how to solve the problem instead of his/her knowledge of using a calculator. These students may also need a "talking" graphic calculator.

To enhance your learning, you must develop an "Individual College Learning/Testing Plan." This plan needs to be developed with your counselor or learning specialist for students with disabilities. The plan includes your academic strengths, weaknesses, services, learning accommodations, testing accommodations and semester goals. This plan can be part of your confidential student file. Figure 19 (Individual College Learning/Testing Plan), on the facing page, is a sample college plan for a fall semester.

Traumatic Brain Injury

Since each student with a TBI is different, a cognitive and achievement assessment is needed. If you were assessed after the TBI, talk to your counselor for students with disabilities about your strengths and weaknesses.

If you were not assessed after your TBI, ask your counselor for students with disabilities for a cognitive and math achievement assessment. Without knowing your processing strengths and weaknesses, it is very difficult to suggest effective study skills and appropriate accommodations.

The study skills you used in the past may no longer be effective. For example, if your TBI affected your visual memory, you may no longer be a visual learner. Use the results from your processing tests to find out how you now learn best.

Some of the study skills on which you must focus are self-monitoring, organization and structure. Some examples of self-monitoring are thinking before you talk and making a "to do" list every day. Your "to do" list can include attending classes, doing home work, attending tutorial sessions, counselor appointments and personal needs.

Review your "to do" list several times a day and cross off the items as you complete them. Add items to your "to do" list as they come up.

Keep an appointment book with important dates such as the times

---

Figure 19
Completed Individual College Learning/Testing Plan

Semester: Fall 1997

A. Student Information:

Name: Joe College
Disability: Learning disability

B. Services:

Tutoring three times a week in math
Word processing lessons twice a week

C. Courses:

Elementary Algebra
History
English
Psychology
Math Study Skills

D. Learning Accommodations:

Note-taker
Tape recorder

E. Testing Accommodations:

Double test time
Enlarged tests
Calculator

Continued on the next page.

Figure 19
Completed Individual College Learning/Testing Plan, *continued*

F. Disability Information:

Strengths: language skills, short-term and long-term memory, reasoning skill, excellent motivation.

Weaknesses: visual processing skills, visual memory skills, reading speed, math study skills, math, time management.

G. Semester Goals:

1. Joe will obtain a 3.0 GPA.

2. Joe will make a "B" in elementary algebra.

3. Joe will improve his math study skills weakness by working on the suggestions from the computerized Math Study Skills Evaluation.

4. Joe will set up a study schedule and each week will complete a study-goals sheet.

5. Joe will attend his tutorial sessions.

6. Joe will use his learning and testing accommodations.

7. Joe will see his math instructor every two weeks and after each major test.

for your final exams. Improving organization can include getting your materials ready for each class and setting up a good study environment.

Structure can help you by getting you into the routine of having a tutorial session, doing homework, going to the library and eating lunch at the same time each day. These suggestions can reduce some of the effects of your memory problems. Form more specific study skills, review the study skills mentioned in the LD section and consult your counselor for students with disabilities.

The learning and testing accommodations you will need are based on your processing deficits. You will have processing deficits similar to those of students with LD. Most likely you will have a short-term memory processing deficit. This means you will need a note-taker for your class and a tape recorder to record the important aspects of the tutorial sessions.

Other processing deficits may be found in working memory, long-term memory and fluid reasoning. Use the accommodations suggestions for students with LD as a guide for your accommodations. Make sure you talk to your counselor for students with disabilities about your appropriate accommodation and make up an Individual College Learning/Testing Plan.

Attention Deficit Disorder

Students with ADD need to be more careful when scheduling their classes than do other students. These students should carry a reduced course load, usually 12 or fewer credit hours. Time should be left vacant before and after each class to allow for review and extra time for test-taking.

Students with ADD should schedule math classes during the time when they are most alert and, if they are on medication, when it is most effective. They should also try to have instructors who use multi-sensory teaching techniques. Once you find a instructor with whom you are successful, you must, at any cost, continue with that instructor in your next math classes.

It is very important to establish the math level to begin the math-course sequence. This level should be one at which the students feel challenged but also can experience success in math. Make sure to take the placement or diagnostic test at your college or university. Being at the correct level is a must for students with ADD. In fact, if there is a question about where you should be placed, always take the lower class. It is better to make an "A" in the lower-level math class than a "D" or "F" in the higher-level course.

Another skill essential for students with ADD is time management. Students should learn to use weekly, term and yearly calendars to plan the semester. Make sure to complete all the time management suggestions in Chapter 4, "How to Fit Study Time Into a Busy Schedule," p. 97. Keep a pocket calendar for your semester schedule and appointments. Check this calendar daily.

Schedule short math study periods each day rather than studying math for longer periods every other day. Large assignments should be divided into smaller, more manageable chunks. Review previously learned material before each new study session.

Students with ADD need to develop ways to make every minute of class time beneficial. First of all, you must sit in The Golden Triangle of Success (P. 113) near the front of the room and away from the windows and doors. This area will have the fewest distractions.

Try to understand the reason for each step in a procedure instead of just following the instructor. If it becomes too difficult for you to understand the instructor and take notes at the same time, ask your counselor for a note-taker and use a tape recorder. Ask questions in class as soon as you get confused. If all your questions cannot be answered, schedule an appointment with the instructor or review the video tapes for that concept.

Students with ADD should use color and action as much as possible during study time. Colored pencils may be used to highlight key concepts or to distinguish grouping symbols, operations and like terms. When working word problems, ADD students should draw pictures or diagrams to aid in organizing the information. Key words should be colored. Students should then follow the suggestions in this text for solving word problems.

Students with ADD should take tests in a distraction-free environment and be given extended test time. Most students with ADD need extended test time because of their uncontrollable internal and external distractions. These distractions, which do not constantly occur with non-disabled students, cause short-term memory problems.

It takes students with short-term memory problems longer to read the test and work each problem step. The extra test time is given to test the students' math knowledge instead of their disability. Review the learning and testing accommodations, in the LD section, for students with short-term memory.

Since you will be taking the test in a separate area, make sure you get any special instructions that will be told to the class. Talk to your counselor for students with disabilities for any additional accommodations that you might need and develop an Individual College Learning/Testing Plan.

A final strategy for students with ADD is self-monitoring. You should keep a record of your own progress, recording each grade received and determining your grade average after each test. You should schedule regular appointments with the instructor to discuss your strengths, weaknesses and grades. Self-monitoring can also be applied to homework, test-taking and progress toward your educational and personal goals. Use the additional self-monitoring techniques suggested in the TBI section.

ADD is one of the least understood disabilities and is challenged more often by instructors and administrators. Talk to the counselor for students with disabilities about ways to discuss your disability with math instructors.

Students with ADD can be successful in math, but skills which may not be essential to other students become essential to the student with ADD. Remember these skills are time management, self-monitoring, math study skills and test-taking skills. By knowing how to study, you can learn more in less time. Once success is achieved, it becomes its own motivator.

How to make an "Individual College Learning/Testing Plan"

Some students with disabilities may have difficulty developing an "Individual College Learning/Testing Plan" (ICLTP). The purpose of the plan is to set semester goals and to obtain the appropriate accommodations for each of your classes.

Defining the ICLTP

The ICLTP is similar to the Individual Education Plan (IEP) that you may have had in middle school or high school. In most cases, the IEP was conducted through a meeting with your parents, teachers and the special education instructor. The IEP detailed the amount and type of help you would receive from your teachers and special education instructor. You had little input on the IEP.

Colleges' and universities' accommodations are based on different laws than the IEP. College and university accommodations are based on federal laws, such as Section 504 and the Americans with Disabilities Act. You need to be prepared to give the Section 504 coordinator or Disability Support Services office personal documentation of your disability.

Included in that documentation should be information on your processing problems and suggested accommodations. You also should be able to explain your disability learning problems to the Disability Support Services office personal. Unlike the IEP meeting, you will be the one who will have the input on the ICLTP.

The counselor cannot contact your previous teacher or even your parents without your permission. In fact the counselor cannot tell your instructor that you are disabled without your permission.

What You Should Do to Get Help

Several weeks before classes begin, meet with your counselor for students with disabilities to discuss your needs. If you are currently enrolled as a disabled student and do not have accommodations, contact the counselor for disabled students as soon as possible. Use Figure 19 (Individual College Learning/Testing Plan), pp. 257-258, as a basic plan to obtain services from that office.

Discuss with the counselor your learning problems and the accommodations which you had in middle school or high school. If your disability was documented after high school, discuss the psychoeducational report results with your counselor.

The psychoeducational report should contain information on your learning weaknesses and suggestions for accommodations. If your report does not have these two areas, request the report to be rewritten to include these items.

How to Help Yourself

Using Figure 19 (Individual College Learning/Testing Plan), pp. 257-258, as a guide for developing your own ICTLP. Ask about the different types of services offered to students with disabilities. You can request tutoring services even though it is not required under Section 504 or the ADA. However, most colleges do offer tutoring through the Disability Support Service office. Also ask for services related to computer training or priority registration. Use all the services that will improve your learning skills.

List all the courses that you will take or are taking. Indicate the courses in which you will need accommodation(s). In most cases, you will not need accommodations in all your classes. Indicate the learning accommodations which will be needed in each course.

The learning accommodations may be the same for each course (such a note-taker), except lab courses. Next, list the testing accommodations you are requesting. In most cases, students can receive double time and testing in a private room. Ask for additional testing accommodations based on your previous IEP or documentation.

Completing Your ICTLP

Section F of the ICTLP is for disability information. List your learning strengths and learning weaknesses based on your information and input from your counselor. This will help the counselor remember how best to help you and to work with you in the future.

Section G concerns developing semester goals. You may need a second meeting with your counselor to establish these goals. These goals should relate to your expected grades, study skills improvement, study schedule and using the services offered by the Disability Support Service office.

Developing an Individual College Learning/Testing Plan is in your best interest. The information is confidential and is kept in a separate file, usually in the Disability Support Service office. Information about your disability cannot be placed in your regular college file or in any way be indicated on your transcripts. Only the instructors from which you request accommodations will know that your are disabled, but they cannot obtain any other information about your disability without your permission.

I would suggest that you meet with these instructors and discuss how your disability affects learning. Even with the confidentiality guarantees, some students, based on previous bad experiences, do not want anyone to know about their disability. These students put off requesting their accommodations until they are failing the course, which is too late.

Do not wait! Request your accommodations now, so you will have the best chance to pass math and your other subjects.

Summary

- Colleges and universities have more students with disabilities attending math classes.

- Students with disabilities need to know their processing deficits and strengths to become successful in math.

- Math study skills and appropriate learning and testing accommodations can compensate for part of their disability.

- The student's accommodations must be matched to their processing deficits and must be made on an individual basis.

- Additional study skills techniques, such as color coding and the Note Card Check System (p. 252), will improve learning.

- Students with disabilities must understand how their disability affects their learning and work closely with their instructor and counselor.

- Students with LD may have difficulty in visual processing speed and visual discrimination that affects math learning.

- Students with TBI will, in most cases, have problems in math due to short-term memory problems and abstract reasoning.

- Students with ADD may have difficulty with math due to concentration and organizational problems.

- For these students to improve their math learning, they will need to develop an Individual College Learning/Testing Plan (including procedures to improve their math study skills and the appropriate learning and testing accommodations), pp. 257-58. This plan needs to be developed before class begins each semester or at least during the first week of classes.

— Students with disabilities must educate themselves about their disabilities and become self-advocates.

— Even though these suggestions are for students with disabilities, students who are not disabled may benefit from many of these study-skills suggestions.

   Remember: Students with disabilities must understand their math-learning weaknesses and consult their counselor for appropriate learning and testing accommodations.

Assignment for Chapter 11

1. In your own words, what is the definition of a learning disability?

2. Who are some famous people with learning disabilities?

3. Explain two reasons why a learning disability causes math-learning problems.

4. Describe two special study skills for students with LD.

5. How are accommodations determined for students with LD, TBI and ADD?

6. What are some problems students with TBI will have learning math, and how can they be accommodated?

7. What are some problems students with ADD will have learning math, and how can they be accommodated?

8. For students with disabilities, describe your Individual College Learning/Testing Plan (see pp. 257-58).

9. For students without disabilities, what are two new study-skills suggestions you could use to improve your math learning?

Glossary

Accommodations – Academic adjustments which allow students with disabilities the same access to information as non-disabled students. These accommodations may be in the form of note-takers, audiocassette recorders, large-print tests and extended test times. Accommodations are given to measure the students' knowledge instead of their disabilities.

Acronym – A memory technique in which on or more words make up the first letter of each of the words in the information you wish to remember. For example, ROY G BIV is an acronym representing the colors of the rainbow – **r**ed, **o**range, **y**ellow, **g**reen, **b**lue, **i**ndigo and **v**iolet.

Adjunct Math Faculty – Part-time math instructors who are employed to teach one to three courses. They usually do not have office hours and usually cannot meet with students who need extra help.

Affective Characteristics – Characteristics students possess which affect their course grades, excluding cognitive entry skills. Some of these characteristics are anxiety, study habits, study attitudes, self-concept, motivation and test-taking skills.

Americans with Disabilities Act (ADA) – Signed into law in 1990, it is intended to provide equal opportunities for persons with disabilities. The ADA did not replace Section 504, but it expanded the provisions of Section 504 to include private businesses. The ADA protects the same student population against discrimination as in Section 504.

Analytic Learning Style – A cognitive learning style that urges students to know the facts and what the experts say. They learn best by thinking about ideas and developing concepts. More interested in ideas than in people.

Assessment Instruments – Surveys use to determine your math anxiety, study skills, attitude, locus of control and procrastination. Other assessments instrument can measure your math knowledge or predict success in your math courses.

Association Learning – A memory technique used to relate new information to be learned to old information you already know.

Attention Deficit Disorder (ADD) – A neurological-based disability that effects a student's ability to concentrate. ADD students are usually impulsive, easily distracted and may be hyperactive. Students with ADD have difficulty concentrating for short periods of time and have difficulty understanding sequential material.

Auditory Processing – The ability to understand the difference between the sounds of words. This is not a problem with hearing or listening, but a central auditory processing disorder in which certain words are drowned out by other noises or some words sound like different words. The student, especially in large lecture halls, has difficulty understanding the lecture due to background noises.

Cognitive Learning Style – The style that focuses on how a student acts on the information that they received. Students act on information differently based on the four cognitive styles: Innovative, Analytic, Common Sense and Dynamic.

Collaborative Learning – A teaching technique that requires students to get into groups to work on math problems. Each group usually has four members. The instructor works with one group while members in other groups teach themselves. The groups are sometimes graded on their answers.

Common Sense Learning Style – A cognitive learning style in which students want to know how the concept or idea works. They want to know the common use of the material. They test their own theories to learn the material. They want to know how the information will help them in real life.

Conditioned Response – A habit developed by repeating the same behavior over and over again.

Cue-Controlled Relaxation – A relaxation-response technique by which students can relax themselves by repeating certain cue words to themselves. A good example of this is that, upon hearing certain old songs (cue words), your feelings (emotions) often change.

Discussion of Rules – Part of the modified, three-column note-taking system in which students write down their lecture notes and important rules used to solve the problems presented in class.

Distributive Learning – A learning system in which you spread your homework on a particular subject over several days instead of trying to do it all at one time. For example, studying for about an hour and taking a five- to 10-minute break before continuing to study.

Dynamic Learning Style – A cognitive learning style by which students want to know unusual possibilities which can be achieved with the information. They learn best by self-discovery and using trial-and-error solutions. They like to carry out their plans to complete their tasks.

Effective Listening – A behavior in which you sit in the least distracting area of the classroom and become actively involved in the lecture.

External Student – A student who believes that he/she is not in control of his/her own life and that he/she cannot obtain his/her desired goals, like making a good grade in math. External students blame their teachers, parents – anyone and anything, except themselves – for their failures.

Fear of Failure – A personal defense mechanism by which a student puts off doing his/her homework so he/she may have an excuse when he/she does poorly in or fails the course. Thus the student's real ability is never measured.

Glossary of Terms – A section in the back of a notebook developed by the student which contains a list of key words or concepts and their meanings.

Highlighting – Underlining important material in the textbook or in your notes with a felt-tip pen.

Individual College Learning/Testing Plan – Developed with a counselor, it lists the type of learning and testing accommodations for students with disabilities. The plan is based on the students processing strengths and weaknesses. It also contains the student's goals for the semester.

Innovative Learning Style – A cognitive learning style in which students want to see personal meaning in learning the material. They want concrete examples and want to reflect it back to themselves. They are interested in people and are imaginative.

Internal Student – A student who believes that he/she is in control of his/her own life and can obtain his/her desired goals – like making a good grade in math.

Learned Helplessness – A lack of motivation due to repeated tries to obtain a goal (like passing math) but failing to obtain that goal. An attitude of "Why try?" develops because of numerous previous failures.

Learning Disability – A neurological condition that blocks learning certain types of information. These students have average- to above-average intelligence. The effects of learning disabilities include problems in processing information into the brain and/or understanding the information once it reaches the brain. Learning disabilities cannot be cured; however, some types of learning disabilities can, to a certain extent, be accommodated.

Learning Modalities – A learning style that focuses on how students prefer input information. The three ways most students input in information is visually, auditorially or kinesthetically. Visual learners prefer to see the information they need to learn. Auditory learners prefer to hear the information they need to learn. Kinesthetic learning prefer to touch and feel the information they need to learn.

Learning Resource Center (LRC) – A location on campus, sometimes in the library, that has educational materials and tutors. These centers could help students in different subjects or could be designed to help students in certain subjects, such as math.

Learning Style – The preferred way a student learns information. Learning styles can focus on how students prefer to input information, which are modalities (visual, auditory, kinesthetic). Learning styles also include how student processes information, which are cognitive styles. Students may have different learning styles for different subjects.

Locus of Control – The belief that one is in control of his/her own life, or that other people or events are controlling his/her life.

Long-Term Memory – Part of the memory chain that retains unlimited information for long periods of time and is considered to be a person's total knowledge.

Long-Term Retrieval – See working memory.

Manipulatives – Hand-help objects that can be used to represent arithmetic and algebra concepts. Students can handle these objects, concretely, to obtain a better understanding how math formulas are used.

Mass Learning – Bunching up all your learning periods at once. For example, trying to complete all your math homework for the last two weeks in one night. This technique is an ineffective way of learning math.

Math Achievement Characteristics – Characteristics which students possess that affect their grades, such as previous math knowledge, level of test anxiety, study habits, motivation and test-taking skills.

Math Attitude – How a student feels about math. Some students feel positive about math and like the subject. Other students have a negative attitude toward math and do not like it. Your attitude toward math could influence how you study and who you blame for poor grades.

Math Glossary – The description of math words that you put in your own words. It usually consists of the bold-print words printed in the textbook.

Math Knowledge – The level of arithmetic and algebra a student possesses when first entering a particular math course. Students with good math knowledge have a chance to be successful in their math class, while students with poor math knowledge will likely fail their math course.

Math Reform – The movement among math faculty and math organizations to change the way math is taught and graded. Math reform came about due to the poor performance of math students and a need to teach all students math. This reform is not a new type of math; rather, it is a different way to teach math using new technology, collaborative learning, learning styles, study skills and real-life math problems.

Memory – The process of receiving information through your senses, storing the information in your mind and recalling the information for later use.

Memory Output – The ability to recall information in verbal or written form. This information is recalled during tests to answer questions or can be used in working memory to solve problems. Memory output can be blocked by high test anxiety.

Mental Picture – A memory technique in which you visualize the information you wish to learn by closing your eyes and forming an image of the material in your mind's eye.

Mnemonic Device – A memory technique in which you develop easy-to-remember words, phrases and rhymes, and relate them to difficult-to-remember concepts.

Negative Math Attitude – A dislike for math that might have been formed during elementary or middle school. Negative math attitude is closely related to math anxiety. It can lead to students' putting off math study, and it is an excuse for failing math.

Negative Self-Talk – One of the causes of test anxiety. It is what you tell yourself during a math test that decreases your confidence. For example, during the test, telling yourself that you are going to fail it and there is nothing you can do about it.

Note Cards – 3x5" index cards; students write important concepts on the front of the card and an explanation of the concepts on the back.

Note-Taking Cues — Signals given by instructors to their classes which indicate that the material they are presenting is important enough that the students may be tested on it. Notes should be taken on this material.

Number Sense — The same as common sense but used with numbers. The ability to look at your answer to see if it makes sense. For example, in a rate-and-distance problem, the rate of the car cannot be 2,000 miles per hour. Number sense is also the ability to estimate the answer to a problem without using your "mental blackboard."

Perfectionist — One who expects to be perfect at everything he/she does, including making an "A" in math when it may be, for him/her, virtually impossible.

Positive Self-Talk — A way to decrease test anxiety. It is what you tell yourself during a math test that increases your confidence. For example, telling yourself that you are going to work as hard as you can to pass the test.

Prerequisite — A course that is the preparation for the course in which you are trying to enroll. Math prerequisite courses have the math background required to be successful in the next math course.

Problem Log — A list of the problems the instructor worked out in class. In a separate section of your notebook, keep the list of problems your instructor worked out — without the solutions. Use these problems as a pretest before taking each major test.

Procrastination — A personal defense mechanism in which one puts off doing certain tasks, like homework, in order to protect one's self-esteem.

Quality of Instruction — The effectiveness of math instructors when presenting material to students in the classroom and math lab. This effectiveness depends on the course textbook, class atmosphere, teaching style, extra teaching aids (videos, audio tapes) and other assistance.

Reasoning – That part of the brain that is used to understand the abstractness of math. Abstract learning is understanding the logical sequence of math problem solving. Reasoning can also be called non-verbal learning.

Relaxation Response – A learned technique that decreases emotional anxiety and/or disruptive thought patterns, allowing you to think more clearly.

Reworking Notes – The process of reviewing class notes to rewrite illegible words, fill in the gaps and add key words or ideas.

Rule of Three – Used to check the solution of a problem when using a graphing calculator. Check your solution three ways: either algebraically, numerically or graphically, depending on how the problems was presented.

Section 504 – A federal law that prohibits discrimination based on disabilities at all post-secondary educational programs receiving federal funds. Section 504 includes students with LD, TBI and ADD. Section 504 suggests academic adjustments (accommodations) which will assist in compensating for students' disabilities.

Sensory Register – The first part of the memory chain that receives the information through your senses (seeing, hearing, feeling and touching).

Self-Monitoring – The ability to record your own behavior to evaluate the progress of obtaining a certain goal. For example, recording the amount of hours studied per week and comparing it to your goal.

Sequential Learning Pattern – A learning pattern in which one concept builds on the next concept. The ability to learn new math concepts is based on your previous math knowledge. Not knowing under-lying math concepts causes gaps in learning, which often results in lower future test scores and even failure.

Short-Term Memory – The second part of the memory chain which allows you to remember facts for immediate use. These facts are soon forgotten.

Skimming – The first step in reading a textbook. It involves over-viewing the chapter to get a general understanding of the material.

Study Buddy – A student who is usually taking the same math course as yourself, and whom you can call for help when you have difficulty doing your math homework.

Test Anxiety – A learned emotional response or thought pattern that disrupts or delays a student's ability to recall information needed to solve the problems.

Test Analysis – A process of reviewing previous tests for consistently misread directions, careless errors, application, test-taking and study errors to help prevent their future occurrences.

Time Management – A process of gaining control over time to help you obtain your desired goals. Using a study schedule is an example of gaining control over time.

Tools of Your Trade – Any material you require to begin studying.

Traumatic Brain Injury (TBI) – Caused by a blow to the head, which causes the person to lose consciousness, creates memory loss and con-fusion, resulting in a neurological deficit. TBI is a disability that usu-ally causes short-term memory problems and, sometimes, problems associated to reasoning.

Video Tapes – Tapes on math concepts that are made by the textbook publisher, math instructors or bought commercially. The tapes can be previewed before class or after class to better understand the concept.

Visual Discrimination – The ability to tell the difference between numbers and letters that look similar. It is not a problem with "seeing" the numbers or letters. Students with visual discrimination problems will have difficulty telling the difference between a "b "and a "d" or between an "x" and a "+" sign. Students with major visual discrimination problems are dyslexic.

Visual Processing – The ability to remember visual information and recognize the difference between similar symbols. Students with poor visual processing may have poor visual memory and/or not be able to tell the difference between similar mathematical symbols. These student may copy material incorrectly off the board and miscopy algebra steps.

Visual Processing Speed – How fast a student can visually read information without making mistakes. The information being read can be numbers or symbols. Students with slow visual processing speed cannot quickly write down numbers, and they are poor note-takers.

Weekly Study Goals – The amount of time scheduled for studying each of your subjects over the period of a week.

Working Memory – Part of the learning process that is used to understand information long enough to place it into long-term memory. The second part of working memory is the space in the brain that is used to recall information from long-term memory and used to work on problems. Working memory is limited, meaning that only a certain numbers of processes/calculations can be done at the same time. Working memory can be compared to the amount of RAM in a computer.

Reference A

10 Steps to Improving Your Study Skills

Improving your study skills can be the great educational equalizer. Effective studying is the one element guaranteed to produce good grades in school. But it is ironic that students are almost never taught how to study – *effectively* – in school.

> **Example:** An important part of studying is note-taking, yet few students receive any instruction in this skill. At best, you are told simply, "You had better take notes," but not given any advice on what to record or how to use the material as a learning tool.

Fortunately, reliable data on how to study does exist. It has been scientifically demonstrated that one method of note-taking is better than another and that there are routes to more effective reviewing, memorizing and textbook reading as well. The following are 10 proven steps you can take to improve your study habits. I guarantee that if you *really* use them, your grades will improve.

1. **Behavior modification can work for you.** Use the association learning concept. Attempt, as nearly as possible, to study the same subject at the same time in the same place each day. You will find that, after a very short while, when you get to that time and place, you are automatically in the subject "groove."

    Train your brain to think math on a time-place cue, and it will no longer take you 10 minutes a day to get in the math mood. Not only will you save the time and emotional energy you once needed to psych yourself up to do math, or whatever else, you will also remember more of what you are studying.

    After studying, reinforce yourself by doing something you *want* to do (watch television, go to a party). Experts know that positive reinforcement of a behavior (such as studying) will increase its frequency and duration.

2. **Do not study more than an hour at a time without taking a break.** In fact, if you are doing straight memorization, do not spend more than 20 to 30 minutes at a time. Here is the rationale behind taking such small bites out of study time.

    First, when you are under an imposed time restriction, you use the time more efficiently. Have you noticed how much studying you manage to cram into the day before big exams? That is why it is called "cramming."

    Second, psychologists say that you learn best in short takes. In fact, studies have shown that as much is learned in four one-hour sessions distributed over four days as in one marathon six-hour session during one day. That is because, between study times, while you are sleeping or eating or reading a novel, your mind subconsciously works on absorbing what you have learned. So it counts as study time, too.

    Keep in mind when you are memorizing, whether it is math formulas or a foreign language or names and dates, that you are doing much more real learning more quickly than when you are reading a social studies text or an English essay.

    The specialists say you will get your most effective studying done if you take a 10-minute break every hour. In fact, some good students study 45 minutes to an hour, and they take a five- to 10-minute break. The break is considered your reward and improves your learning over the next hour.

Dr. Walter Pauk, Director of the Reading and Study Center at Cornell University, suggests you take that short break when- ever you feel you need one. That way, you will not waste your time away by clock-watching and anticipating your break.

Another technique for keeping your mind from wandering while studying is to begin with your most boring subject – or your hardest one – and work toward the easiest and/or the one you like best. Thus, your reward for studying the most boring or hardest is studying the subject you like best. Try it; it works.

3. **Separate the study of subjects that are alike.** Brain waves are like radio waves. If there is not enough space between input, you get interference. The more similar the kinds of learning taking place, the more interference. So, separate your study periods for courses with similar subject matter. Follow your studying of math with an hour of Spanish or history, not chemistry or statistics.

4. **Do not study when you are tired.** Psychologists have found that everyone has a certain time of day when he or she gets sleepy. Do not try to study during that time (but do not go to sleep either – it hardly ever refreshes). Instead, schedule some physical activ- ity for that period, such as recreation. If you have a pile of school- work, use that time to sort your notes or clear up your desk and get your books together or study with a friend.

5. **Prepare for your class at the best time.** If it is a lecture course, do your studying soon after class; if it is a course in which stu- dents are called on to recite or answer questions, study before class. After the lecture, you can review and organize your notes. Before the recitation classes, you can spend your time memoriz- ing, brushing up on your facts and preparing questions about the

previous recitation. Question-posing is a good technique for helping the material sink in and for pinpointing areas in which you need more work.

6. **Use the best note-taking system for you.** Quite a bit of research had been done on note-taking, and one system has emerged as the best. Use 8½-by-11-inch loose-leaf paper and write on just one side. (This may seem wasteful, but it is one time when economizing is secondary.) Take the time to rule your page as follows:

    a. If the course is one in which lecture and text are closely related, use the 2-3-3-2 technique: Make columns of two inches down the left-hand side for *recall clues*, three inches in the middle for *lecture notes* and three inches on the right side for *text notes*. Leave a two-inch space across the bottom of the page for your own *observations and conclusions*. See Figure 20 (Three-Column Note-Taking System) on the next page.

    b. If it is a course where the lectures and the reading are not closely related, use separate pages for class notes and reading notes, following the 2-5-1 technique: Two inches at left for *recall clues*, five in the middle for *lecture notes* and an inch at the right for *observations and conclusions*. (After a while, you will not need to draw actual lines.)

    You have most likely taken your lecture notes in the form that evolved during your years of schooling. You have also probably evolved your own shorthand system, such as using a "g" for all "-ing" endings, an ampersand (&) for "and," and abbreviations for many words (e.g., govt. for government and evaptn. for evaporation).

    The *recall clue column* is the key to higher marks. As soon as possible after you have written your notes, take the time to read them over — not studying them, just *reading* them. Check

Figure 20
Three-Column Note-Taking System

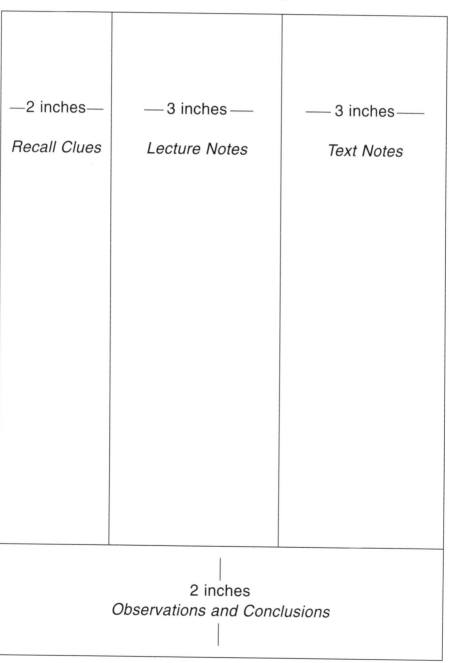

right away, while it is all still fresh, to see whether you have left out anything important or put down anything incorrectly, and make changes.

After reviewing what you have written, set down *recall clue words* to the topics in your notes. These clue words should not repeat information but should designate or label the kind of information that is in your notes. They are the kind of clues you would put on "crib sheets."

Example: To remember the information contained so far in this section on note-taking, you need just the following clues: 8½-by-11, loose-leaf, one side: 2-3-3-2 or 2-5-1. As you can see, they are simply memory cues to use later on in your actual studying.

Dr. Robert A. Palmatier, Assistant Professor of Reading Education at the University of Georgia, suggests that you study for tests in the following manner: Take out your loose-leaf pages and shift them around so the order makes the most sense for studying.

Choose the first page and cover up the notes portion, leaving visible just the clues. See if you can recall the notes that go with the clues. As you get a page right, set it aside.

If you are going to be taking a short-answer test, shuffle your note pages so that they are out of order. (That is why it is important to use just one side of the paper.)

"This approach provides for learning without the support of logical sequence," Dr. Palmatier says, "thus, closely approximating the actual pattern in which the information must be recalled."

If you are going to be taking an essay test, you can safely predict that "those areas on which the most notes are taken will most often be the areas on which essay questions will be based."

The beauty of the "recall clue word" note-taking method is that it provides a painless way to do the one thing proved to help you remember what you have learned — actively thinking about the notes and making logical sense of them in your mind. If, instead, you just keep going over your recorded notes, not only will you get bored, but you will be trying to memorize in the worst way possible.

7. **Memorize actively, not passively.** Researchers have found that the worst way to memorize — the way that takes the most time and results in the least retention — is to simply read something over and over again. If that is the way you memorize, forget it. Instead, use as many of your senses as possible.

    Try to *visualize* in concrete terms, to get a picture in your head. In Addition to sight use sound: *Say* the words out loud and *listen* to yourself saying them.

    Use association: Relate the fact to be learned to something personally significant or find a logical tie-in.

> **Examples:** When memorizing dates, relate them to important events, the dates of which you already know. Use mnemonics: For example, the phrase "Every good boy does fine," is used for remembering the names of the musical notes on the lines of the treble clef. Use acronyms, like OK4R, which is the key to remembering the steps in the reading method outlined in number 8, below.

8. **Read and study at the same time.** It really takes less time in the long run! Read with a purpose. Instead of just starting at the beginning and reading through to the end, you will complete the assignment much faster and remember much more if you first take the time to follow the OK4R method devised by Dr. Walter Pauk:

    O   **Overview.** Read the title, the introductory and summarizing paragraphs and all the headings included in the reading material. Then you will have a general idea of what topics will be discussed.

    K   **Key Ideas.** Go back and skim the text for the key ideas (usually found in the first sentence of each paragraph). Also read the italics and bold type, bulleted sections, itemizations, pictures and tables.

R1 **Read** *your assignment from beginning to end.* You will be able to do it quickly, because you already know where the author is going and what he/she is trying to prove.

R2 **Recall.** Put aside the text and say or write, in a few key words or sentences, the major points of what you have read. It has been proven that most forgetting takes place immediately after initial learning. Dr. Pauk says, "One minute spent in immediate recall nearly doubles retention of that piece of data!"

R3 **Reflect.** The previous step helps to fix the material in your mind. To cement it there forever, relate it to other knowledge; find relationships and significance for what you have read.

R4 **Review.** This step does not take place right away. It should be done for the next short quiz, and then again for later tests throughout the term. Several reviews will make that knowledge indelibly yours.

9. **Make up a color and sign system for text and notes.** For your text, Dr. Palmatier suggests:

Red for main ideas

Blue for dates and numbers

Yellow for supporting facts. Circles, boxes, stars and checks in the margins can also be utilized to make reviewing easy.

Make your own glossary of the words and concepts you do not know.

In your notebook, underline, star or otherwise mark the ideas which your teacher tells you are important: Thoughts to which you are told you will be coming back later, items which you are warned to be common mistakes. Watch for the words – such as *therefore* and *in essence* – which tell you what is being summarized. Always record examples. In fact, in such subjects as math, your notes should consist mainly of your teacher's examples.

Pay close attention in your note-taking until the last minute of class time. Often, a teacher gets sidetracked and runs out of time. He/she may jam up to a half-hour's content into the last five or 10 minutes of a lecture. Get down that packed-few minutes' worth. If necessary, stay on after class to get it all down.

10. **Do not buy underlined textbooks.** Of course, if the book does not belong to you, you will not be underling at all. But if you underline, do it sparingly. The best underlining is not as productive as the worst note-taking.

Over-underlining is a common fault of students; only the key words in a paragraph should be underlined. It should be done in ink or felt-tip highlighter, and it should be done only after you have finished the "OK" part of your OK4R reading.

If you are buying your books secondhand, never buy one that has already been underlined. You may tend to rely on it, and you have no idea whether the hand that helped the pencil got an "A" or a "F" in the course! If, due to availability or finances, you have to buy an underlined textbook, mark it in a different color.

Research has proven that it is not how much time you study that counts but how well you study during that time. In fact, in at least one survey, students who studied more than 35 hours a week came out with poorer grades than those who studied less.

Remember: Use your study time wisely, and you too will come out ahead.

Reference B

Suggestions to Students for Improving Math Study Skills

1. View a math study skills video tape.

2. Learn how to relax before tests are taken.

3. Use a good math note-taking system.

4. Spend as much time on math homework as needed.

5. Complete your most difficult homework assignments first. Usually, this means math homework.

6. Read ahead in the math textbook and prepare questions for the instructor.

7. For each chapter, prepare your own list of math vocabulary words.

8. Find a study buddy and set up group study times.

9. Develop practice tests and time yourself while taking them.

10. Read ahead in your textbook and make an informal outline.

11. For practice, do all the example problems in the text.

12. While doing homework, write down questions for the instructor/tutor.

13. Be aware of the time allotted while taking a math test.

14. Make sure you attend every math class.

15. Schedule a study period after your math class.

16. Review video tapes of different math classes. For difficult topics, review the video tapes before going to class.

17. Verbalize (silently) problems the instructor writes on the board. Solve the problem or silently verbalize each solution step.

18. Interview instructors before actually signing up for the class. Compare your learning style to their instructional style.

19. Make note cards to remind yourself how to solve various types of math problems.

20. Get help early in the semester before you get too lost in the course.

21. For understanding, recite back the materials you have read in the math textbook.

22. Take notes on how to solve difficult problems.

23. Copy all the information that is written on the board.

24. The book, *How To Solve Word problems* by Mildred Johnson (McGraw-Hill, 1992), is recommended for use by students having difficulty with word problems.

25. Do math homework every day.

26. If you miss a class, ask your instructors for permission to attend the same course that is taught at a different time or day.

   Remember: You are held responsible for material covered in classes that you have missed.

Reference C

Student Learning-Style Information

Visual Numerical Learners

These students learn math best by seeing it written. If you are a visual numerical learner, you may learn best by following these suggestions:

1. Using worksheets, workbooks and tests as additional references.

2. Studying a variety of written material, such as additional math texts.

3. Playing games with and being involved in activities with visible, printed number problems.

4. Using visually orientated computer programs and CD-ROMs.

5. Taking good notes and reviewing them every other day.

6. Reworking your notes using suggestions from *Winning at Math*.

7. Review someone's notes while comparing them to your notes.

8. Visualizing numbers and formulas, in detail.

9. Viewing checked-out video cassette tapes from the math lab or LRC.

10. Making 3x5 note (flash) cards, in color.

11. Using different colors of ink to emphasize different parts of the math formula.

12. Using a highlighter or felt-tip pen to underline important material in your notes or textbook.

13. Asking your tutor to *show you* how to do the problems instead of *telling you* how to do the problems.

14. Writing down each problem step the tutor tells you to do. Highlight the important steps or concepts which cause you difficulty.

15. Asking for additional handouts on the math materials.

Auditory Numerical Learners

If you are an auditory numerical learner, you may learn math best by *hearing* information and *discussing* how to do the problems. You may learn best by:

1. Playing auditory games which involve math.

2. Saying numbers to yourself or moving your lips as you read problems.

3. Listening to math audio tapes which may be found in the math lab.

4. Tape recording your class and playing it back while reading your notes.

5. Reading aloud any written explanations.

6. Explaining to your tutor how to work the math problem.

7. Making sure all important facts are spoken aloud.

8. Remembering important facts by auditory repetition.

9. Studying in an area with a low noise level.

10. Reading math problems out loud and trying solutions verbally and sub-verbally as you talk yourself through the problem.

11. Recording directions to difficult math problems on audio tape and referring to them when solving a specific type of math problem.

12. Having your tutor explain how to work the problems instead of just showing you how to solve them.

13. Finding an instructor who explains, in detail, how to solve math problems.

14. Using a study group to discuss with other students how to solve math problems.

15. Recording math laws and rules in your own words, by chapters, and listening to them every other day (auditory highlighting).

Social Group Learners

Social group learners may learn best in a group setting and in collaborative learning by discussing information. If you are a social group learner, you may learn most efficiently by:

1. Studying math, English or your other subjects in a study group.

2. Signing up for math course sections which use cooperative learning (learning in small groups).

3. Signing up for courses which have group discussion such as philosophy or logic.

4. Obtaining help in the math lab or other labs where you can work in group situations.

5. Watching math video cassette tapes with a group and discussing the subject matter.

6. Listening to audio cassette tapes on the lecture and discussing them with the group.

7. Obtaining several "study buddies" so you can discuss with them the steps to solving math problems.

Social Individual Learner

The social individual learner may learn best alone. If you are a social individual learner, you may learn best by:

1. Studying math, English or other subjects alone.

2. Utilizing video cassette tapes or auditory tapes to learn by yourself.

3. Preparing individual questions for your tutor or instructor.

4. Obtaining individual help from the math lab or hiring your own tutor.

5. Setting up a study schedule and study area so you will not be bothered by other people.

6. Studying in the library or in some other private, quiet place.

7. Using group study times only as a way to ask questions, obtain information and take pretests on your subject material.

Tactile Concrete Learners

Tactile concrete learners who are also called *kinesthetic* learners tend to learn best when they can concretely manipulate the information to be learned. This learning style creates a problem with math learning because math is more abstract than concrete.

Most math instructors are visual abstract learners and have difficulty teaching math tactilely. Ask for the math instructors and tutors who give the most practical examples and who may even "act out" the math problems.

Ask your math lab supervisor if he/she has any manipulative materials or models which can be used to learn math. If the math lab does not have any manipulative materials, you may have to ask your instructor to help you make up some.

If you are a tactile concrete learner you will probably learn most efficiently by hands-on learning. Learning is most effective when physical involvement with manipulation is combined with sight and sound. You may learn best by:

1. Trying to get involved with at least one other student, tutor or instructor.

2. Obtaining diagrams, objects or manipulatives and incorporating activities such as drawing and writing into your study time. You may also enhance your learning by doing some type of physical activity, such as walking, while learning.

3. Trying to use your hands and body to "act out" a solution. For example, you may "become" the car in a rate-and-distance word problem.

Visual Language Learners

Visual language learners best learn courses in areas such as English, humanities, speech and social science by seeing the information. If you are a visual language learner, you may learn best by:

1. Using books, pamphlets and other written materials.

2. Reading the material before class.

3. Taking good notes using a color-coding system.

4. Getting a copy of someone else's notes and comparing them to your notes.

5. Making outlines of your notes and memorizing the outline.

6. Employing reading techniques which include occasionally stopping and visualizing the scene or information.

7. Using a highlighter or felt-tip pen to underline important materials in your textbook, study guides or notes.

8. Making 3x5 note cards to learn new terms and vocabulary words.

9. Visualizing written information in detail.

10. Keeping your work area free of visual distractions.

11. Planning what your are going to say ahead of time by writing down your thoughts and listing questions.

12. Using a computer to study different subject areas.

13. Having your tutor or instructor show you how to do your homework instead of telling you how to do it.

14. Using visual computer programs to improve your writing and reading skills.

15. Checking out video cassette tapes on subject areas you are currently learning.

Auditory Language Learners

Auditory language learners best learn courses in areas such as English, humanities, speech and social science by hearing the information. If you are an auditory language learner, you may learn best by:

1. Using a tape recorder in class and later listening to the tapes.

2. Participating in small group discussions.

3. Reading aloud or moving your lips as you memorize information.

4. Listening to lectures and repeating back to yourself the important information.

5. Participating in class discussions.

6. Listening to audio tapes on the subjects you are taking.

7. Explaining to other students the difficult information in your subject areas.

8. Setting up appointments to discuss your course materials with your instructors.

9. Improving your spelling by using the phonetics approach.

10. Trying to remember important information by auditory repetition.

11. Working in areas with low noise levels.

12. Having your tutor explain in various ways the correct answers to

your homework assignments.

13. Repeating back to your tutor, in your own words, the correct answers to your homework.

14. Recording important subject information in your own words and listening to the information every other day (auditory highlighting).

15. Trying to find instructors who go into great verbal detail when lecturing.

Reference D

10 Ways to Improve Your Memory

Strategies to use in studying information to be remembered:

1. **Space study/practice** – Spacing study periods is more effective than learning material all at once. Long study periods do not allow an opportunity to consolidate what you have learned. Time and spacing varies: Four one-hour periods result in better recall of material than one four-hour session. Students should study in the evening in short spaced periods, go to bed and get up in the morning and review what was learned the night before.

2. **Active recitation** – As you read and learn information, frequently stop and repeat to yourself what you learned. Put into your own words what you have just learned. This focuses your attention on the material at hand, and it gives you repeated practice in retrieving information that has been stored. It will help you to recall (remembering on your own) rather than just recognizing the information stored.

If it helps you, write the facts down in outline form at the same time that you are reciting the information out loud.

Sometimes it helps to study with someone else. They can ask you questions about the information that you just read or studied. You repeat out loud to them the important facts learned.

3. **Overlearn the material** – Continue to practice beyond the point of bare mastery. Even though you are able to recall it once, continue to practice over and over again. This will increase the amount of material that you will later be able to remember.

4. **Recall** – Recount something learned.
   Recognition – Identify something learned.
   Relearning – How fast one can learn material that was previously learned. Relearning is sometimes called the "savings method," whereby you review material learned earlier in the quarter to refresh your memory rather than cramming all at one time. The more you review during the unit or quarter, the less time you will have to spend going over old material for the exam or evaluation.

5. **Improve external memory** – This refers to all physical devices that help you remember: lists, memos, diaries and alarm clocks. Many of us are either too lazy or too proud to make the best use of such help: We forget to perform a chore because we felt we did not need to jot it down. One handy form of external memory is the deliberately misplaced object that reminds you to do something.

6. **Use chunking** – This means grouping several items of information into one piece that is easy to remember as a single item. We recall an acronym like UNICEF as a single name, not as six letters. To cue us in to remember all of the Great Lakes, many of us use HOMES: Huron, Ontario, Michigan, Erie and Superior.

   Psychologist Laird Cermak, author of *Improving Your Memory* (Johns Hopkins, 1994), urges you to make up your own chunks. His example: For a picnic, you need milk, soda, beer, salami, bologna, liverwurst, napkins, paper cups and paper plates. If you do not have a pencil handy, that is a lot to remember. Yet you can make it easy. There are three drinks, three meats, and three paper goods; use the first letter of each category – d, m, p – to make a word: damp (bad for picnics). Remember that, and you will recall the categories and the items in each.

7. **Use mediation** – This means attaching the items of a list to some easily remembered "mediating" device, such as the jingle most adults use to recall the lengths of the months: "30 days hath September . . . "

Make up your own mediators. To remember all the things to take care of when going away for a weekend, I listed: Water plants, throw out spoilables in refrigerator, turn on telephone answering machine, lower thermostat, lock windows, put out garbage and lock doors. From the first letter of each item, I made up the sentence, "Peter Rabbit takes turns with gourmet dinners." Ridiculous, but easy to remember.

8. **Use associations** – Visual images are one effective form of association. To remember names, think of a visual link between a person's name and some facial feature. For instance: You have just met a Mr. Clausen, who has bushy eyebrows. When you try to recall his name, you see his eyebrows, then remember the claw tearing at them and – *a ha!* – Clausen!

9. **Relive the moment** – Studies have shown that sensory impressions are associated in memory to what we are learning, and they later help remind us of what we learned. If you are trying to recall a name or fact, picture the place in which you learned it, the people around you at the time, even the feeling of the seat in which you sat; your chance of remembering it will be greatly increased.

If you are trying to remember where you lost something, mentally retrace your steps. "Ah!" you may suddenly say, seeing the scene in your mind's eye, "I put the parcel on the empty chair next to me in the restaurant when the waiter handed me the menu."

10. **Use mnemonic pegboards** — Performers who remember scores of names called out by people in their audiences do not have unusual memories; they have previously memorized a set of words or images to which they mentally attach the names. Anyone can do it.

First, memorize these ten "pegwords" (since they rhyme with the numbers one to 10, it is easy):

one – bun
two – shoe
three – tree
four – door
five –hive
six – sticks
seven – heaven
eight – gate
nine – line
ten – hen

Now, make up a list of 10 other words and number them. Link each one to the pegword with the same number by means of an image. Suppose your first word is bowl; picture a bun lying inside a bowl. If your second word is desk, picture a shoe parked on a desk. A minute should be enough for all 10. You will be amazed at how effortlessly – and for how long – you can recall the whole list.

Reference E

Stress

There are many definitions of personal stress. *Webster's Dictionary* describes stress as being "Tense: Strained exertion: As the stress of war affects many people."

Kenneth Lamott, in this book *Escape from Stress* (Berkley Publishing Corp., 1975), defines this condition as not only the well-publicized "executive stress" but the results of a significant disruption in one's environment, which can lead to a stress-response syndrome. Stress can be induced by chronic, minor frustrations, such as being late to class or a day late paying bills. Psychiatrist Ainsley Mears describes stress as a beginning to be looked at as a health problem in today's anxiety-producing world.

Stress, no matter how one defines it, will lead to chronic or acute physical and psychological disabilities. Such disabilities lead to distress, and the result is mental and/or bodily suffering. Some of the mental stress problems can be explained by the fight-or-flight response, which was established as a survival mechanism in the early development of mankind.

When our ancestors were faced with a threat, hormones were automatically released into their bloodstream, and the autonomic nervous system prepared their bodies for immediate action. During fight or flight, the body metabolizes "stress hormones," and at the close of an ordeal, a person is exhausted — but no longer under stress.

The fight-or-flight response is still active and essential in today's life. Survival in warfare, dangerous sports and everyday driving depends on it. Yet, numerous stress events in our lives can activate the fight-or-flight response. The response could be a supervisor asking if you have finished a report or your wife asking when you will be home.

As you face these stressful situations, your autonomic nervous system releases a "stress hormone" which increases our blood pressure, heart rate, rate of breathing and metabolism, preparing you for the event. If this process is repeated often enough over a period of time, you may become anxious, uptight or short tempered, leading to mental stress and physical problems.

Physical problems manifest themselves as high blood pressure, peptic ulcers and various heart diseases. Uncontrolled stress will lead to a shorter, more miserable life.

"What can I do to control the fight-or-flight response?" you ask. The answer is nothing: It is an autonomic system, and you have no control over it. But you can control the *effects* of the response.

As mentioned, a side effect of the fight-or-flight response is tension. Sports of all kinds are frequently used to combat stress. So are music, painting and other hobbies. A trip to the country or the beach will relieve the pressures of the day. Alcohol and other drugs-tranquilizers, barbiturates and marijuana will also reduce tension, but they have dangerous side effects. There are other ways to reduce tension not mentioned above, but most of them fall into two areas: Chemical and situational.

A third alternative is what Dr. Benson calls, "An innate protective mechanism against 'overstress,' which will counter the effects of the fight-or-flight response."

What Dr. Benson is referring to is the *relaxation response*. The relaxation response can be induced by various exercises or rituals used over generations. Some examples are: Zen, Yoga, T.M., self-hypnosis, biofeedback, progressive relaxation, auto-genic training and The Benson Method. The last three examples mentioned above will be the process taught to the group.

Relaxation Response

When anxiety or tension is brought on by a stressful event, the quality of feelings may be different depending upon the event. People will experience this anxiety in different ways such as feeling a "knot" in the stomach or having their "mind racing in circles." The person who has a tense stomach is experiencing *somatic anxiety* (anxiety of the body) and the other person is experiencing *cognitive anxiety* (anxiety of the mind — worry). There can, however, be variations of both types of anxiety in one person, though one is usually dominant.

Given that these are the two basic forms of anxiety, one type of relaxation response should have more effect on cognitive versus somatic anxiety and vice versa. The literature shows that there seems to be a definite effect on cognitive or somatic anxiety depending on which relaxation response you use.

In general, progressive relaxation reduces somatic anxiety. The Benson Method reduces cognitive anxiety. Although autogenic training seems to affect mostly somatic anxiety, it also affects cognitive anxiety. Further explanations of each relaxation response follow.

Progressive Relaxation

Progressive relaxation was developed by Dr. Jacobson in 1938 and is probably the most widely used relaxation technique today. This relaxation technique entails systematically tensing and relaxing about 16 different muscle groups.

The therapist instructs the subject. On the instructor's command the subject will tense one muscle group and then relax it. Upon relaxing, the subject will focus attention on the same muscle group again. This sequence of tensing and relaxing one muscle group will be systematically applied to each major muscle group of the body. The result will be somatic relaxation.

Autogenic Training

Autogenic training was first developed by J. H. Schultz in the 1900s. This procedure is a passive way to achieve somatic relaxation.

The subject is placed in a comfortable position, perhaps lying on a mat or in a reclining chair. Verbal instruction are given to relax different parts of the body. Suggestions of heaviness and warmth are given to the subject to induce a feeling of relaxation.

Once all parts of the body are relaxed subject imagine a tranquil scene. Typical scenes are beaches, the forests and listening to music. This positive imagery helps the subject sink deeper into relaxation and reduces cognitive anxiety. The results are reductions in both somatic and cognitive anxiety.

The Benson Method

The most recent technique to induce the relaxation response was produced by Dr. Herbert Benson. This technique is called "The Benson Method" and was developed as an alternative to Transcendental Meditation. There are four basic components necessary to bring forth the proper response: A quite environment, a comfortable position, a mental device and a passive attitude

To explain this technique, a subject is first directed into a quiet room. The subject is then asked to close his/her eyes and relax all muscles. The subject is asked to become aware of his/her breathing, which is through the nose.

As the subject breathes out, the word "One" is silently repeated. For example, breathe in . . . out. "One" In . . . out. "One," etc. The subject breathes easily and naturally and continues this process for 10 to 20 minutes. During that time, breathing will automatically decrease but should not be forced.

If any distracting thoughts occur, they should be allowed to pass through the subject's mind, but the mind should not dwell on them. Instead, return to repeating "One." After the time duration, the subject opens his/her eyes and moves parts of his/her body to come out of the meditation. As mentioned before, this relaxation response reduces cognitive anxiety, and to some extent, somatic anxiety.

Daily Use of Relaxation

At some time during any given day, we all feel ourselves getting tense and tired. We are sure that the next remark we hear might cause us to lose control. There is not enough time to enact The Benson Method or progressive relaxation. What, then?

The answer is twofold: First, do a short relaxation response (which might be palming), and, second, initiate differential relaxation as one of several preventive measures.

Palming

Palming is done with the eyes closed and covered with the palms of the hands, with elbows resting on a desk or table. To avoid exerting any pressure on your eyeballs, the lower part of your palms should rest upon your cheek bones and your fingers should rest upon your forehead.

Once the light is blocked out, you can start visualizing some relaxing scenes developed during autogenic training. This should reduce your anxiety to a coping level.

Differential Relaxation

Differential relaxation should be practiced as a preventative measure. Differential relaxation is a way to relax while being active. It involves learning to differentiate between those muscles necessary to perform a task and those which are unnecessary.

The muscles not necessary to perform a task should be relaxed to conserve nervous and muscle energy. For instance, while sitting and reading a book, you can relax your legs while maintaining minimal tension in only the upper part of your body.

As you go through your daily activities, be aware of needless tension in some parts of your body and relax them.

The two techniques mentioned are only some of the ways you can control tension in your daily life. The best way to control tension is to be prepared for it by practicing the relaxation responses best suited for your lifestyle.

Remember: Prevention is the best policy in coping with stress.

Reference F

Studying for Exams

What to know before you start to study:

1. What type of test is it?

    a. Objective – multiple choice, true-false, matching or a combination.
    b. Essay – short or long answer, or sentence completion.
    c. Problem solving.
    d. Combination of the above.

2. What material is to be covered?

3. How many questions (approximately)?

4. What is the time limit?

    If the information above is not given by the instruction when he/

she announces the test, *ASK*. This information is valuable to the way you study. Also, ask the instructor for old exams you can use for your review.

Studying

1. Be sure you have read all the material to be covered and have all the lecture notes *before* you begin your serious studying.

2. Plan what you will study and when you will study it.

3. Each review session should be limited to one hour. Take short breaks of five to 10 minutes between hourly sessions.

4. Try to predict exam questions. If it will be essay, try writing out the answer to your predicted questions.

5. Study in a group *only* if everyone has read the material. You do not gain much when you must "tutor" someone else or if the other students are not prepared.

6. Prepare summary sheets to study and eliminate rereading the textbook.

7. Review for objective tests by concentrating on detail and memorizing facts, such as names, dates, formulas and definitions (know a little bit about a lot).

8. Review for essay tests by concentrating on concepts, principles, theories and relationships (know a lot about a little bit).

9. For problem-solving tests, work examples of each type of problem. Work them from memory until you get stuck. Study your guide problem and begin working it again from memory, from the

beginning. Do this until you can work the entire problem without referring to your notes.

10. On the day of the test, do not learn any new materials. It can interfere with the knowledge you have already learned.

11. Try not to discuss the test with other students while you are waiting to begin. If you have studied, you do not need to be flustered by others making confusing remarks.

12. Try to consciously make yourself relax before the test begins.

13. After the test is over, forget it! Do not discuss it and do not look for answers you might have missed. Concentrate on your next exam.

14. Keep in good physical condition by not ignoring food and/or sleep requirements.

Reference G

Answering an Essay Test

Essay tests can have on them the following types of questions: Short or long answer, fill in the blank, and sentence completion. Use the following suggestions to help you with essay-type tests:

1. **Make a brief survey of the entire test.** Read every question and the directions. Plan to answer the least difficult questions first, saving the most difficult for last.

2. **Set a time schedule and periodically check your progress** (to maintain proper speed). With six questions to answer in 60 minutes you should allow a maximum of 10 minutes per questions. If your 10 minutes passes and you have not finished the question, continue to the next one and come back to the other one later. Do not sacrifice any question for another.

3. **Read the question carefully.** Underline key words: <u>List</u>, <u>compare</u>, <u>WWII</u>, political <u>and</u> social, art <u>or</u> music, etc. As you read,

jot down the points that occur to you beside that question.

4. **Organize a brief outline of the main ideas you want to present.**
    Place a check mark alongside each major idea and number them
    in order of presentation in your answer. Do not spend too much
    time on the outline.

5. **When you answer, always rephrase the question.**

> **Example:** Explain Pavlov's theory of conditioning. Answer: Pavlov's
> theory of conditioning is based on . . .

The remainder of the answer is devoted to support by giving
dates, examples, stating relationships, causes, effects and research.

6. **Present material that reflects the *grader's* personal or profes-
    sional biases.** Further, stick to the material covered in the reading
    or lecture, and answer the question within the frame of reference.

7. **If you do not understand what the instructor is looking for, write
    down how you interpreted the question and answer it.**

8. **If time does not permit a complete answer, use an outline form.**

9. **Write something for every question.** When you "go blank," start
    writing all the ideas you remember from your studying — one of
    them is bound to be close!

10. **In sentence-completion items, remember never to leave a space
    blank.** When in doubt — GUESS. Make use of grammar to help
    decide the correct answer. Make the completed statement logi-
    cally consistent.

11. **If you have some time remaining, read over your answer.** You
    can frequently add other ideas which may come to mind. You can

at least correct misspelled words or insert words to complete an idea.

12. **Sometimes, before you even read the questions, you might write some facts and formulas you have memorized on the back of the test.**

Reference H

Taking an Objective Test

Objective tests include those with multiple-choice, true-false or matching questions. Use the following suggestions to help you take an objective test:

1. **Before you start taking the test, preview the entire test** – survey to find how many questions there are and of what type. Set a time limit so that you will have at least five minutes at the end to recheck your test.

2. **Read the directions, carefully, making sure you understand exactly what is expected.**

3. **Find out if you are penalized for guessing.** If not, always guess and don not leave any unanswered questions.

4. **Carefully read each question; underline key words.**

5. **Anticipate the answer and look for it.** Read all the alternatives before answering.

6. **Do not read into questions what is not there.**

7. **When your anticipated answer is not one of the options, discard it and systematically concentrate on the given ones.**

8. **When two or more options look correct, compare them with each other.** Study them to find what makes them different. Choose the more encompassing option unless the question requires a specific answer.

9. **Pass over the difficult or debatable questions on your first reading and come back to them after completing those about which you were sure.**

10. **Use information from other questions.**

11. **In all questions, especially the true-false type, look for specific determiners.** Words such as "rarely," "usually," "sometimes," and "seldom" allow for exceptions; "never," "always," "no," and "all" indicate no exceptions.

12. **Mark statements true only if they are true without exception.** If any part of the statement is false, the whole statement is marked as such.

13. **Stay in one column of a matching test.** Usually it will be the column with the definition. Work backward to find the word or symbol that matches it. Be sure to find out if the answers can be used more than once.

14. **If you know you made an error, change your first answer.** If it is just a guess, keep your first impression.

Index

A

F

Fall semester — 41, 256
family responsibilities — 102
fear of failure — 84, 87-88, 93, 272
fear of success — 84, 87, 89
fight-or-flight response — 309-10
final exam — 75, 105, 141, 220, 233-34, 236-37, 256
foreign language — 34, 35, 36, 48, 50, 123, 282

G

getting stuck — 25, 126, 134, 140, 143, 153, 210, 220, 224, 230
grade goals — 100, 104, 110
glossary of terms — 272
grading system — 39, 40, 41, 50
guessing — 48, 219, 224-25, 320, 323-24

H

helplessness — 84, 86, 87, 93, 206, 273
highlighting — 136, 138, 192-93, 249, 251, 260, 272, 296, 202
homework —25, 37-38, 43, 66, 87, 89-90, 102, 108, 114, 125-28, 131,
        133-35, 138-44, 147-49, 152-57, 164, 170-71, 187, 191-92, 210-
        12, 219-21, 235, 237, 245, 247-51, 259, 261, 271-72, 274, 276,
        278, 289-91, 301, 303
homework goals — 138

I

J

L

Q

quality of instruction — 52, 56, 276

R

reading ahead — 137-38, 289-90
reading material — 114, 135, 154, 286
reading skills — 66, 67, 79, 301
reading math textbooks — 114, 135, 154, 286, 289-90
reasoning — 182, 187, 190, 241, 243, 246-47, 255, 258-59, 265,
    277-78
rebelling against authority — 90
recitation — 122, 191, 192, 193, 283-84, 290, 305
Relaxation — 103, 211-15, 217-18, 271, 277, 310-14
relaxation response — 212, 214, 271, 277, 310-14
Rework — 129
reworking notes — 112, 122-23, 129, 164, 277
rule of three — 277
Rush — 230
rushing — 87, 284

S

SAT — 54, 221, 240, 265, 277, 307
Schultz, J. H. — 312
scratch paper — 194, 225-26, 231
Section 504 — 254, 262-63, 270, 277
self-concept — 52-53, 270
self-hypnosis — 311
self-monitoring — 256, 261, 277
self-talk — 209, 212, 215-18

T

V

W

Y

Z

Figures Index

Bibliography

Bloom, B. (1976). *Human Characteristics and School Learning.* New York: McGraw-Hill Book Company.

Fox, Lynn H. (1977). *Women and Mathematics: Research perspective for change.* Washington, D.C.: National Institute of Education.

Ieffingwell, B. J. (1980). "Reduction of test anxiety in students enrolled in mathematics courses: Practical solutions for counselors." Atlanta, Georgia: A presentation at the Annual Convention of the American Personnel and Guidance Association. (ERIC Document Reproduction Services No. ED 195-001)

Kolb, D. (1985). *Learning-Style Inventory.* McBer and Company.

McCarthy, B. (1987). *The 4-mat system: Teaching to learning styles with right/left mode techniques.* Barrington, IL: Excel.

Nolting, P. D. (1987). *How to Reduce Test Anxiety*, an audio cassette tape. Pompano Beach, FL: Academic Success Press.

Nolting, P. D. (1987). *How to Ace Tests*, an audio cassette tape. Pompano Beach, FL: Academic Success Press.

Nolting, P. D. (1989). *Strategy Cards for Higher Grades*. Pompano Beach, FL: Academic Success Press.

Nolting, P. D. (1991). *The Effects of Counseling and Study Skills Training on Mathematics Academic Achievement*. Bradenton, FL: Academic Success Press.

Nolting, P. D. (1994). *Improving Mathematics Study Skills and Test-Taking Skills*. Boston, MA: D. C. Heath, Inc.

Nowicki, S., & Duke, M. (1974). "A locus of control scale for non-college as well as college adults." *Journal of Personality Assessment*. No. 38, pp. 136-137.

Nowicki, S., & Duke, M. (1978). "Examination of counseling variables within a social learning framework." *Journal of Counseling Psychology*. No. 8, pp. 1-7.

Richardson. F. C., and Suinn, R. M. (1973). "A comparison of traditional systematic desensitization, accelerated mass desensitization, and mathematics anxiety." *Behavior Therapy*. No. 4, pp. 212-218.

Tobias, S. (1978). "Who's afraid of math and why?" *Atlantic Monthly*. September, pp. 63-65.

Suggested Reading

Adams, J. D. (1980). *Understanding and Managing Stress.* San Diego, CA: University Associates, Inc.

Ellis, D. B. (1985). *Becoming a Master Student.* Rapid City, SD: College Survival, Inc.

Johnson, M. (1992). *How to Solve Word Problems in Algebra: A Solved Problems Approach.* New York, NY: McGraw-Hill Book Company.

Mason, L. T. (1986). *Guide to Stress Reduction.* Berkeley, CA: Celestial Arts.

Murphy, S. T. (1992). *On Being L. D.* New York, NY: Teachers College Press.

Nadeau, K. G. (1994). *Survival Guide for College Students with ADD or LD.* New York, NY: Magination Press.

Nolting, P. D. (1987). *How to Reduce Test Anxiety.* Bradenton, FL: Academic Success Press. Audio cassette tape.

Nolting, P. D. (1987). *How to Ace Tests.* Bradenton, FL: Academic Success Press. Audio cassette tape.

Nolting, P. D. (1989). *Strategy Cards for Higher Grades.* Bradenton, FL: Academic Success Press. Twenty 3x5-inch cards.

Nolting, P. D. (1991). *Math and the Learning Disabled Student: A Practical Guide for Accommodations.* Bradenton, FL: Academic Success Press.

Nolting, P. D. (1993). *Math and Students with Learning Disabilities: A Practical Guide to Course Substitutions.* Bradenton, FL: Academic Success Press.

Nolting, P. D. (1994). *Improving Mathematics Studying and Test-Taking Skills.* Boston, MA: D. C. Heath and Co. Videotape.

Station, T. F. (1977). *How to Study.* Nashville, TN: How To Study.

Palladino, C. (1994). *Developing Self-Esteem for Students.* Menlo Park, CA: Crisp Publications, Inc.

Weiss, L. (1992). *Attention Deficit Disorder in Adults.* Dallas, TX: Taylor Publishing Company.

About the Author

Dr. Paul D. Nolting

Over the past 15 years, Learning Specialist Dr. Paul Nolting has helped thousands of students improve their math learning and obtain better grades. Dr. Nolting is a national expert in assessing math learning problems – from study skills to learning disabilities – and developing effective learning strategies and testing accommodations.

Dr. Nolting is also a nationally recognized consultant and trainer of math study skills and of learning and testing accommodations for students with learning disabilities. He has conducted national training grant workshops on math learning for the Association on Higher Education and Disabilities and for a two-year, University of Wyoming, 1991 Training Grant of Basic Skills.

Dr. Nolting has conducted numerous national conference workshops on math learning for the National Developmental Education Association, the National Council of Educational Opportunity Association and the American Mathematical Association of Two-Year Colleges. He is a consultant for the American College Test (ACT) and the Texas Higher Education Coordinating Board. He is also a consultant for Houghton Mifflin Faculty Development Programs conducting workshops on math study skills for math faculty. His text, *Winning at Math: Your Guide To Learning Mathematics Through Successful Study Skills* is used throughout the United States, Canada and the world as the definitive text for math study skills.

Dr. Nolting has consulted with numerous universities and colleges. Some of the universities with which Dr. Nolting has consulted are the University of Massachusetts, the University of Colorado-Bolder, Texas Tech. University, Black Hills State University, Tennessee Tech. University and the University of Connecticut.

Some of the colleges with which he has consulted are San Antonio College, St. Louis Community College, J. Sargeant Reynolds College, Montgomery College, Broward Community College, Miami-Dade Community College, Northeast State Technical Community College, Landmark College, Denver Community College, Valencia Community College and Pensacola Junior College. Dr. Nolting has consulted with over 75 colleges, universities and high schools over the last 15 years.

Dr. Nolting holds a Ph.D. degree in Education from the University of South Florida. His Ph.D. dissertation was "The Effects of Counseling and Study Skills Training on Mathematics Academic Achievement." He is an adjunct instructor for the University of South Florida, teaching Assessment and Appraisal courses at the graduate level.

His book, *Winning at Math: Your Guide to Learning Mathematics Through Successful Study Skills* was selected Book of the Year by the National Association of Independent Publishers. "The strength of the book is the way the writer leads a reluctant student through a course from choosing a teacher to preparing for the final examination," says *Mathematics Teacher*, a publication of the National Council of Teachers of Mathematics.

His two audio cassettes, *How to Reduce Test Anxiety* and *How to Ace Tests* were also winners of awards in the National Association of Independent Publishers' competition. "Dr. Nolting," says *Publishers's Report*, "is an innovative and outstanding educator and learning specialist."

A key speaker at numerous regional and national education conferences and conventions, Dr. Nolting has been widely acclaimed for his ability to communicate with faculty and students on the subject of improving math learning.

Other Books and Materials by Academic Success Press, Inc., and Dr. Paul Nolting

Audio Cassette Tapes

Nolting, P. D. *How to Ace Tests.* Bradenton, FL: Academic Success Press, 1987. Audio cassette tape. $9.95.

Nolting, P. D. *How to Reduce Test Anxiety.* Bradenton, FL: Academic Success Press, 1987. Audio cassette tape. $9.95.

Books

Nolting, P. D. *The Effects of Counseling and Study Skills Training on Mathematics Academic Achievement.* Bradenton, FL: Academic Success Press, 1990. $19.95.

Nolting, P. D. *How to Develop Your Own Math Study Skills Workshop and Course.* Bradenton, FL: Academic Success Press, 1991. $12.95.

Nolting, P. D. *Math and The Learning Disabled Student: A Practical Guide For Accommodations.* Bradenton, FL: Academic Success Press, 1991. $14.95.

Nolting, P. D. *Math and Students with Learning Disabilities: A Practical Guide to Course Substitutions.* Bradenton, FL: Academic Success, 1993. $19.95.

Card Decks

Nolting, P. D. *Strategy Cards for Higher Grades.* Bradenton, FL: Academic Success Press, 1989. Twenty 3x5-inch cards. $5.95

Notebooks

Nolting, P. D. *Math Study Skills Notebook.* Bradenton, FL: Academic Success Press, 1995. Notebook. $4.95.

Posters

Nolting, P. D. *Ten Steps for Doing Your Math/Science Homework.* Bradenton, FL: Academic Success Press, 1988. Poster. $7.95.

Nolting, P. D. *Ten Steps to Better Test-Taking.* Bradenton, FL: Academic Success Press, 1988. $7.95.

Nolting, P. D. *Six Types of Test-Taking Errors.* Bradenton, FL: Academic Success Press, 1988. Poster. $7.95

Nolting, P. D. *Translating English Terms Into Algebra Symbols.* Bradenton, FL: Academic Success Press, 1988. $7.95.

Savage, B. *Ten Steps For Solving Story Problems.* Bradenton, FL: Academic Success Press. Poster, 1990. $7.95.

Software

Winning at Math — Computer Evaluation Software. Bradenton, FL: Academic Success Press. Software. $99.95.

Video Tapes

Nolting, P D. *Improving Mathematics Studying and Test-Taking Skills.* Boston, MA: D. C. Heath and Co, 1994. Videotape. $59.95.

Ordering Information

For ordering information on *Winning at Math* and other books and learning materials listed on the previous pages, call toll free: 1-800-247-6553. Credit card, checks and purchase orders are accepted.

Shipping charges are $4.00 for the first item and $2.00 for any additional items.

Academic Success Press can be contacted at: PO Box 25002, Box 132, Bradenton, FL 24206.

For a free catalog call (941) 359-2819.

What Students Say

"I thought you'd like to know that I got 100% on my final exam in algebra this semester. That makes ten consecutive 100's I've gotten on math exams since taking your course 'Math Study Habits'.

"Prior to enrollment in your class, I was very apprehensive about math. In fact, I used to count on my fingers and struggled with basic addition. Now, as a result of your help, I have the confidence to complete the prerequisite courses for physical therapy which include triganometry [sic] and physics.

"Not only did your course help me in math, but in all my other subjects as well. I feel 'Math Study Habits' should be required for all entering freshmen.

"Thanks again for all your help."

"I want to thank you for all the valuable information I learned in your Study Habits class the first half of this semester. I don't know if many students come back and tell you of the success they achieved using your techniques, but I want you to know that I got my first 98 on an algebra test and I was thrilled.

"I have been a poor math student from day one and getting a 98 gave me a lot of confidence in my ability. Not learning how to study as a child, I really learned things I wish I'd been exposed to years ago. I still don't spend as much time as I should on homework, but the time I spend is very productive and I look for improvement in all of my classes

"Thank you again."

"During the Fall semester of 1987 I was enrolled in your Study Habits course (REA 1605). Your lectures were both interesting and informative, and the course material a significant factor in reducing test anxiety and ensuring higher grades.

"I am pleased to state that, since completing your course, I have received A's in all subjects (including math, - my most difficult) with the exception of one B in a non-math elective.

"It is my belief that this could not have been accomplished without the knowledge of methods and techniques acquired in your REA 1605, and I would recommend this course to all students wishing to raise their GPA, and particularly those who may need to adjust their study habits and/or having difficulty taking exams."

"I have enjoyed your book *Winning at Math*. The part of reducing test anxiety has helped me a great deal. Thanks for having the visualization procedure. My favorite place is the beach & thinking of the setting sun relaxes me. The only question I have is if you work in other subjects besides math? If so I would be interested in reading those books."

"Translating the English Terms into algebraic symbols is a great idea. I didn't realize how many different terms and symbols they have. It helps to kind of memorize them. They help a lot with word problems."

"*Winning at Math* is a very informative book. I have a different and more positive look at math and life after taking this course. I would highly recommend this class."

"After reading your book, I have tools I may use every day and I thank you."